Successful Development of Green Building Projects

Focusing on office buildings, this book explores how Green Building (GB) development can be managed to achieve successful project outcomes. The book starts by highlighting the special requirements of GBs which distinguish them from traditional buildings. The book then presents a detailed discussion of the success conditions for GB projects. Highlighting 73 success conditions which have been categorised within 20 broad themes, the book reports on the findings from interviews with GB stakeholders from Australia, Hong Kong, Pakistan, Singapore, the United Arab Emirates, and the United Kingdom. The book demonstrates how the complexity, design methodology, and team collaboration prevalent in the delivery of successful GBs set them apart from traditional building projects. The book also demonstrates that success in GB delivery is generally associated with socio-technical conditions.

The research reported in this book will allow project decision-makers such as clients and project team members to consider the wide range of identified success conditions to optimise project performance across its development stages and achieve successful project outcomes. Theoretically, the findings can inform future research focused on GB development, resulting in the more efficient development of GB projects that can reduce the effects of climate change and resource depletion.

Tayyab Ahmad is an assistant professor at the Department of Civil and Architectural Engineering, Qatar University. He has degrees in architectural engineering and construction management. His expertise is in building energy efficiency, sustainable design optimisation, project success frameworks, decision-making frameworks, Building Information Modelling (BIM), and life cycle assessment. His PhD research is about investigating project success in GBs. His research is aimed at improving the practice in Architecture, Engineering, and Construction (AEC) industry to meet the needs of the present without compromising the ability of future generations to meet their own needs.

Spon Research

Publishes a stream of advanced books for built environment researchers and professionals from one of the world's leading publishers. The ISSN for the Spon Research programme is ISSN 1940-7653 and the ISSN for the Spon Research E-book programme is ISSN 1940-8005

Corruption in Infrastructure Procurement
Emmanuel Kingsford Owusu and Albert P. C. Chan

Improving the Performance of Construction Industries for Developing Countries
Programmes, Initiatives, Achievements and Challenges
Edited by Pantaleo D Rwelamila and Rashid Abdul Aziz

Work Stress Induced Chronic Diseases in Construction
Discoveries Using Data Analytics
Imriyas Kamardeen

Life-Cycle Greenhouse Gas Emissions of Commercial Buildings
An Analysis for Green-Building Implementation Using A Green Star Rating System
Cuong N. N. Tran, Vivian W. Y. Tam and Khoa N. Le

Data-driven BIM for Energy Efficient Building Design
Saeed Banihashemi, Hamed Golizadeh and Farzad Pour Rahimian

Successful Development of Green Building Projects
Tayyab Ahmad

BIM and Construction Health and Safety
Uncovering, Adoption and Implementation
Hamed Golizadeh, Saeed Banihashemi, Carol Hon and Robin Drogemuller

Successful Development of Green Building Projects

Tayyab Ahmad

R Routledge
Taylor & Francis Group

LONDON AND NEW YORK

First published 2023
by Routledge
4 Park Square, Milton Park, Abingdon, Oxon OX14 4RN

and by Routledge
605 Third Avenue, New York, NY 10158

Routledge is an imprint of the Taylor & Francis Group, an informa business

British Library Cataloguing-in-Publication Data
A catalogue record for this book is available from the British Library

Library of Congress Cataloging-in-Publication Data
Names: Ahmad, Tayyab, author.
Title: Successful development of green building projects/Tayyab Ahmad.
Description: New York, NY: Routledge, 2023. | Series: Spon research |
Includes bibliographical references and index. |
Identifiers: LCCN 2022060666 | ISBN 9781032345468 (hardback) |
ISBN 9781032345482 (paperback) | ISBN 9781003322740 (ebook)
Subjects: LCSH: Office buildings–Design and construction. | Sustainable
construction. | Structural engineering.
Classification: LCC TH4311.A36 2023 | DDC 690/.5230286–dc23/
eng/20230111
LC record available at https://lccn.loc.gov/2022060666

ISBN: 978-1-032-34546-8 (hbk)
ISBN: 978-1-032-34548-2 (pbk)
ISBN: 978-1-003-32274-0 (ebk)

DOI: 10.1201/9781003322740

Typeset in Sabon LT Std
by KnowledgeWorks Global Ltd.

Contents

Figures

Tables

Foreword

The construction and operation of buildings are responsible for the use of significant quantities of resources, impacts on ecosystems, and pollutants to air, land, and water. As society encroaches on or exceeds planetary boundaries, there is an increasing imperative for our built environment to be in harmony with the earth's natural systems on which we depend.

Whether due to more stringent government regulations, or voluntary aspirations to be more sustainable, building as we have in the past is not an acceptable way to build in the future. In any change process, there is uncertainty and risk, as those involved are forced to replace familiar patterns of activity with new ones. Building green introduces additional measures of success for a project beyond time, cost, and quality. To meet these, additional measures might require different approaches to design and construction, new supply chains, and new materials and products.

Given the urgency of becoming more sustainable as a society, we must then ask the question: how can we accelerate the transition to widespread Green Buildings (GBs)?

This book is part of the answer.

It has taken me 20 years of working in HVAC design, sustainability consultancy, client-side project support, government design review, and advisory and assessing for a GB certification scheme to gain the type of experience that is synthesised in this book.

For those early in their careers, it provides a jump-start to their knowledge base; for those more experienced, it is a prompt for reflection; and for clients, who may not work day-to-day in GBs but have a highly influential role to play, the book provides a checklist of factors to consider.

The author synthesises data from 75 semi-structured interviews of GB professionals from Australia, Hong Kong, Pakistan, Singapore, the United Arab Emirates (UAE), and the United Kingdom. Their experience includes as client, design, engineering, sustainability advisory, facility management, and commissioning. The richness of this dataset is shown via network diagrams and accompanying explanations, while the quotes from interviewees provide relatable examples for the themes and success conditions identified.

I encourage you to read this book and reflect on how you can more effectively contribute to making our built environment more sustainable.

Dr Gerard Healey
Manager, Sustainability Strategy
University of Melbourne

Preface

The building and construction sector tends to consume large amounts of resources. To mitigate these environmental effects, Green Buildings (GBs) stand out as an effective solution. These buildings tend to be much more environment-friendly, socially habitable, and economically affordable in the long term. GB projects are part of the new paradigm in the construction industry as alongside *schedule, cost,* and *quality* parameters, their performance is also determined by *social, economic,* and *environmental* aspects. This subsequently contributes to sustainability in a global context. The construction industry is rapidly approaching a stage where the ability to plan and execute a sustainable project would be a core competency for any top builder. GB projects are becoming increasingly popular in society, and this signifies the need for effective development practices regarding these projects.

Increased attention towards the development practices of GBs is required to boost their increasing rate of construction. The design of project processes is critical to their success or failure (Sanvido & Konchar, 1998), as the research is increasingly showing that the way project teams are composed, the relationships that are formed, the organisation of the project, and the contracts used significantly affect how projects are undertaken and the success of a project (AbulHassan, 2001; Mir & Pinnington, 2014). As a niche segment of the construction industry, GB sector is still in the early stages of its evolution. It is therefore important to focus on the project development practices of GBs. The Architecture, Engineering, and Construction (AEC) industry is beginning to realise that GB projects have additional requirements in their overall development process. These requirements include the use of integrated project delivery, the use of simulation tools, the early involvement of key project parties along with the commitment of a project owner towards sustainability, to name a few. The influence of planners, users/operators, and community services on sustainable projects is much higher than on traditional projects. Furthermore, team collaboration and integration are usually more important for achieving sustainability goals than for traditional goals. Hence, GB projects provide socio-economic and environmental benefits, but they also have special requirements in their development.

If success enabling conditions are ignored during the development process of GB projects, a multitude of problems can result, and project performance may be detrimentally affected. As the requirement for high performance in GB projects escalates, increased optimisation becomes indispensable (Korkmaz, Horman, Molenaar, & Gransberg, 2010). Alongside the organisational and procedural issues, GB project development faces many challenges in the integration of complex building systems, alignment of multi-party interests, use of team experience and knowledge in project development, and the like. The majority of construction projects are still being carried out according to the traditional methods, with short-term solutions preferred in place of long-term solutions, and with the incorporation of managerial solutions, technical approaches, and material selection which can hardly be regarded as innovative green practices (Demaid & Quintas, 2006; Gluch, Gustafsson, & Thuvander, 2009; Hwang & Tan, 2012). This implies that to ensure the increase in number of GB projects and to ease the achievement of sustainability aspirations, more attention is required on the aspects of their development and delivery. There is a need for project development research on the GB market that can guide owners and project teams in their practices to maximise project performance (Korkmaz, Horman, & Riley, 2009). Addressing this need is the aim of this book.

The rationale of the study reported in this book is founded on the shortcomings of previous studies. Previous studies on GB project success have generally ignored the complexity and systems nature of GB project development. Instead of seeing project development holistically, with a vast number of enabling factors and interrelationships among those factors, the previous studies have typically used reductionist approaches in data collection, analysis, and interpretation. Further, previous studies lack an emphasis towards exploratory research design and are excessively reliant on explanatory research design. As a result, these studies lack the attention towards identifying success conditions (conditions enabling successful development of GBs) specifically related to GB projects. Previous studies in this knowledge area are also limited in explaining the success conditions in terms of the theories related to construction. To address the shortcomings of current literature on GB project development and to add depth to the knowledge area, this study considers a large variety of conditions enabling successful development of GBs and the interrelationships among success conditions realistically without oversimplifying those interrelationships. By virtue of this, the study reported in this book uncovers a number of success conditions previously undiscovered. This book addresses two key questions:

Question 1: How do sustainability requirements set GBs apart from non-GBs?

Question 2: What are success conditions in GB projects and how do they interrelate?

This book reports the findings of the PhD research conducted by the author. The scope of this study is limited to the development stage of Green Office projects. The study examines the development process of GB projects, spanning from the project inception to the project hand over. The study highlights a wide variety of success conditions with the potential to influence the performance of GBs not only during the project development stage but also during the use stage. Yet, the operation stage, per se, as well as the end of life, is not included. Data for this study is collected using 75 semi-structured interviews of GB professionals across Australia, Hong Kong, Pakistan, Singapore, the United Arab Emirates (UAE), and the United Kingdom (UK). GB professionals interviewed during data collection include project clients, design consultants, sustainability consultants, facility management (FM) professionals, and commissioning consultants. Acknowledging the difference in sustainable development-related challenges of different building types, the study focuses on Green Office projects to control data variability; help learn from the experience of GB professionals, which is typically in Green Office projects; ensure the comparability of study findings with existing literature since most of previous studies on GB projects have considered Green Office project type; and ensure highly impactful findings since commercial sector is highly energy and resource intensive and can benefit significantly from sustainable development.

This book is divided into two parts. Part I provides an understanding of the attributes setting GBs apart from non-GB projects. Part II provides a detailed account of the social and technical conditions enabling success in GB projects. It also interprets the relative significance of these conditions for project success. The study employs the theoretical lens of Transformation-Flow-Value-generation theory to interpret the significance of success conditions. Implications of the success conditions for theory and practice of GB projects are also addressed in detail in Part II.

Results show that GBs are significantly different from traditional buildings primarily because of their complexity, design methodology, and team collaboration. Good project management tends to associate with high sustainability performance in GBs. To achieve success in GB development, it is necessary to have proficient project teams and clients, high collaboration among the team, highly committed project team and client, rigorous project planning, timely execution of activities, and rigorous process of defining goals. These conditions help realise success in GBs by resulting in value for the project client and reducing the non-value-adding activities during project development.

For practice related to GBs, the findings of this study will allow decision-makers and project teams to consider the wide range of identified conditions to optimise project performance and achieve successful project outcomes. The success conditions can also be considered by GB certification systems in their credit lists to ensure a higher correlation between certification and success. Theoretically, the findings can inform future research focused on GB project development. This will ultimately result in the more efficient development of GB projects that help reduce the effects of climate change as well as resource depletion. Students of built environment can benefit from the findings reported in this study to get a comprehensive understanding of the conditions enabling the success of GB projects.

References

AbulHassan, H. S. (2001). *A framework for applying concurrent engineering principles to the construction industry* (PhD dissertation). Pennsylvania State University. Retrieved from https://elibrary.ru/item.asp?id=5244688

Demaid, A., & Quintas, P. (2006). Knowledge across cultures in the construction industry: Sustainability, innovation and design. *Technovation*, 26(5), 603–610. doi:10.1016/j.technovation.2005.06.003

Gluch, P., Gustafsson, M., & Thuvander, L. (2009). An absorptive capacity model for green innovation and performance in the construction industry. *Construction Management and Economics*, 27(5), 451–464. doi:10.1080/01446190902896645

Hwang, B. G., & Tan, J. S. (2012). Green building project management: Obstacles and solutions for sustainable development. *Sustainable Development*, 20(5), 335–349. doi:10.1002/sd.492

Korkmaz, S., Horman, M., Molenaar, K., & Gransberg, D. (2010). *Influence of project delivery methods on achieving sustainable high performance buildings report on case studies*. Research Sponsored by the Charles Pankow Foundation, DBIA. Retrieved from https://dbia.org/wp-content/uploads/2018/05/Research-Influence-on-Sustainable-High-Performance-Bldgs-Case-Studies.pdf

Korkmaz, S., Horman, M., & Riley, D. (2009). Key attributes of a longitudinal study of green project delivery. Paper presented at the Construction Research Congress, ASCE, Seattle, WA.

Mir, F. A., & Pinnington, A. H. (2014). Exploring the value of project management: Linking project management performance and project success. *International Journal of Project Management*, 32(2), 202–217.

Sanvido, V. E., & Konchar, M. D. (1998). *Project delivery systems: CM at risk, design-build, design-bid-build*. Austin, TX: Construction Industry Institute.

Acknowledgements

After conducting research related to project success for more than three years, I am able to see the factors contributing to my own eventful journey. I am glad that God enabled me to contribute to an endeavour of sustainable development. Thinking of this exciting research journey, I would like to thank:

My parents Nazeer and Ahmad for everything;
Dr Ajibade Aibinu and Dr Andre Stephan for their company, support, training, and motivation;
Prof. Christopher Heywood and Dr Gerard Healey for their valuable feedback;
Dr Jamaluddin Thaheem for his support and advice;
Prof. Albert Chan and Prof. Florence Ling for their guidance;
Joham for her counsel and faith;
Armaghan for his friendship;
My friends Ahmed, Neeraj, Roro, and Waseem at Melbourne School of Design who contributed towards this journey by listening, understanding, and advising;
The University of Melbourne for funding this research.

Particular thanks to industry professionals who consented to participate in this study through semi-structured interviews. I am grateful for the valuable time they spent on this research.

Definitions

Green Buildings: The study will use the term 'Green building' (GB) to refer to the buildings which are developed on the principles of sustainability. These buildings can have the lowest level of sustainability performance as permitted by local regulations in a region, and these buildings can also achieve the maximum potential of sustainability performance as technologically possible. These buildings may and may not be certified by third-party certification systems such as LEED.

Project development: Project development in this study is considered as the GB project stage which spans the whole duration during which the project is conceived and materialised. The way the project is developed plays different direct and indirect roles in defining project outcomes and in achieving those outcomes.

Success conditions: In this study, the factors enabling the achievement of aspired outcomes in GB projects are referred to as success conditions. These conditions are the circumstances, decisions, or events contributing towards project outcomes but do not form the basis of success judgement (Ahmad & Aibinu, 2017; Ika, 2009). For instance, project team collaboration contributes towards the achievement of project outcomes and therefore qualifies as a success condition. In literature, alternate terms have also been used in lieu of success conditions including success factors, critical success factors (CSFs), sustainable performance enabling factors, and Project Delivery Attributes.

Success criteria: The set of principles or standards by which project performance can be evaluated are referred to as success criteria in this study (Ika, 2009; Lim & Mohamed, 1999). In simple terms, the relationship between 'success factors' and 'success criteria' is a relationship of cause factors (for example, the owner's sustainability commitment) with effect factors (such as sustainability performance).

Differentiating conditions: The special characteristics of GB project which differentiate them from their traditional non-green counterparts are referred to as differentiating conditions in this study. Alongside the identification of success conditions, this study also involves the identification of differentiating conditions. Since differentiating conditions are the special characteristics of GB projects, it is probable that many of these conditions are also success conditions. However, instead of making this assumption, the study conducts a separate investigation of differentiating conditions.

Project client: The term 'project client' is used to refer to the party for whom the project procurement takes place and services of the project team are availed. A project client can be the owner of a GB, a developer, an investor, and in some cases also the end-user.

Credits of Green Building certification systems: A credit of a GB certification system (such as LEED) addresses a measure to improve the sustainability performance of a project. A credit defines a clear outcome that the project meets. For instance, 'water metering' as a credit in LEED intends to identify opportunities for water saving. During LEED certification if a project considers water metering credit, it is awarded one point which contributes towards the overall rating of the project.

References

Ahmad, T., & Aibinu, A. A. (2017). Project delivery attributes influencing green building project outcomes: A review and future research directions. *Built Environment Project and Asset Management, 7*(5), 471–489. doi:10.1108/Bepam-11-2016-0066

Ika, L. A. (2009). Project success as a topic in project management journals. *Project Management Journal, 40*(4), 6–19. doi:10.1002/pmj.20137

Lim, C., & Mohamed, M. Z. (1999). Criteria of project success: An exploratory re-examination. *International Journal of Project Management, 17*(4), 243–248.

Abbreviations

AEC	Architecture, Engineering, and Construction
BIM	Building Information Modelling
BOT	Build-Operate-Transfer
BREEAM	Building Research Establishment Environmental Assessment Method
CSF	Critical success factor
ESD	Ecologically sustainable development
GB	Green Building
GBCA	Green Building Council of Australia
HVAC	Heating, ventilation and air conditioning
LEED	Leadership in Energy and Environmental Design
MEP	Mechanical, Electrical, and Plumbing
NABERS	National Australian Built Environment Rating System
PDM	Project delivery method
PM	Project manager
USGBC	US Green Building Council

Part I

Distinct nature of Green Building projects

This part aims to address the question, that is *How do sustainability requirements set Green Buildings apart from non-Green Buildings?*

An in-depth understanding of the successful development in Green Building (GB) projects is not possible without understanding their distinct nature. To investigate the differences among GB and non-GB projects, the findings presented in this part are informed by semi-structured interviews of 25 GB experts. The thematic analysis conducted on interview data helped identify the differentiating conditions of GB projects. Interview participants were mainly based in Australia ($n = 9$) and the UK ($n = 10$) (see Figure PI.1). They were highly experienced building professionals as most of them ($n = 19$; 76%) had a minimum of six years of experience in developing GB projects. These professionals were mostly design consultants ($n = 5$; 20%) and sustainability consultants ($n = 7$; 28%). Up to five interviewees had the experience of acting in multiple roles. Further details regarding the interview participants are provided in Annex A (Section 6.3). Part I has one chapter which provides a detailed account of the conditions differentiating GBs from non-GB projects.

DOI: 10.1201/9781003322740-1

Regions where the Green Building experience of interview participants is mainly based

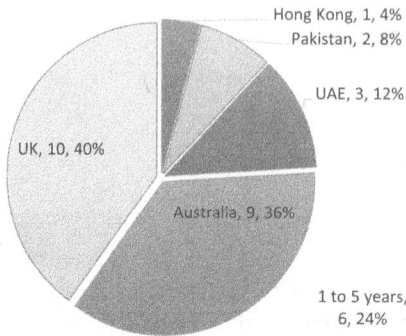

Experience of interview participants in Green Building projects

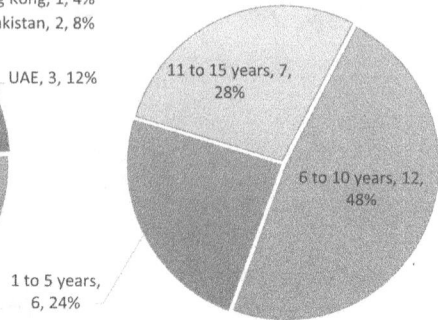

Hong Kong, 1, 4%

Pakistan, 2, 8%

UAE, 3, 12%

UK, 10, 40%

11 to 15 years, 7, 28%

6 to 10 years, 12, 48%

Australia, 9, 36%

1 to 5 years, 6, 24%

Role of interview participants in Green Building projects

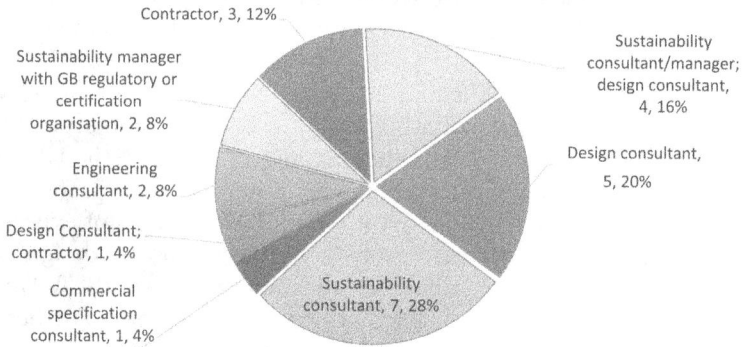

Contractor, 3, 12%

Sustainability manager with GB regulatory or certification organisation, 2, 8%

Engineering consultant, 2, 8%

Design Consultant; contractor, 1, 4%

Commercial specification consultant, 1, 4%

Sustainability consultant/manager; design consultant, 4, 16%

Design consultant, 5, 20%

Sustainability consultant, 7, 28%

Figure Pl.1   Role, experience, and regional belonging of interview participants discussing differentiating conditions.

Chapter 1

What sets Green Buildings apart

1.1 Introduction

> **Who should read this chapter**
>
> Read this chapter:
>
> If you want to know how sustainability requirements set Green Buildings (GBs) apart from non-GBs
>
> If you are interested in understanding the social conditions enabling successful outcomes in GBs, see Chapter 2. If you are interested in understanding the technical conditions enabling successful outcomes in GBs, see Chapter 3.

While investigating 'passive houses', Rohracher (2001) claims that many approaches of traditional building development are rendered obsolete for a passive house project. The traditional planning procedures and the approaches for component certifications, as well as the quality control practices, need to be revised.

To promote the development of better GB projects, it is necessary to understand the differences between GBs and non-GBs, not only in terms of project outcomes but also in terms of the project development process. Such inquiry can help understand why the traditional way of project development may not be used for GB projects. Accordingly, this chapter is about 'identifying conditions which differentiate GBs from non-GB projects'. In this chapter, first, a detailed overview of the differentiating conditions is provided. Subsequently, associations among these conditions are explained and a detailed discussion regarding these conditions is provided.

DOI: 10.1201/9781003322740-2

1.2 Conditions differentiating Green Buildings from their traditional counterparts

Upon analysing interview data, the 29 conditions differentiating GB developments from non-GB developments are grouped into ten themes (see Table 1.1). GB experts strongly agree that GBs differ from non-GBs in terms of design methodology, complexity, and timeliness of project activities. While the identified conditions generally differentiate GBs from non-GB projects, it does not imply that these conditions strictly occur

Table 1.1 Conditions differentiating Green Buildings from non-Green Building projects

Theme	Differentiating conditions	Sub-conditions	Freq.
Budget and time	Increased budget and delivery time requirement		1
Defining project goals	Setting of a detailed sustainability charter or brief	A detailed description of sustainability requirements in the project brief	1
	End-users' involvement in defining project aspirations	Engagement of end-users in the design process	1
Design methodology	Intense design activity	Intense design activity; modelling performed on architectural designs by Mechanical, Electrical, and Plumbing (MEP) consultants; iterative design cycles prior to design finalisation	5
	Level of complexity in project design	Additional design considerations and constraints for consultants	1
	Use of integrated design approach	Integrated design process; focus towards a systems approach (that is, how the systems work together)	4
	Use of innovative design approach	Creative design approach	1
	System and material selection	Diligence in the procurement of building materials	3
	Market survey to identify successful building systems	A continuous search for better building materials and technology	1
	Use of context-oriented design approach	Integration of building within its surroundings	1
	Considerations towards potential building occupants	Integration of users within the building	1
	Use of energy-oriented design approach	Increased focus towards the water, energy usage, and other environmental aspects; on-site energy generation	3
	Life-cycle-based project development approach		1
	Use of holistic design approach		1

(Continued)

Table 1.1 Conditions differentiating Green Buildings from non-Green Building projects (*Continued*)

Theme	Differentiating conditions	Sub-conditions	Freq.
Documentation	Higher documentation activity		2
Flow of project information	Use of effective communication tools and strategies	Meetings to resolve clashes in building services	1
Inspection, monitoring, and control	Monitoring and controlling operational performance of the building	Metering for quick feedback	1
	Execution of commissioning and fine-tuning	Increased commissioning and tuning requirement	1
Team collaboration	Liaison between sustainability consultant and project team	Coordination of sustainability consultant in the design development process	2
	Project team collaboration	Liaison between mechanical engineer, structural engineer, and architect from earlier design stages; integrated project team	2
Team procurement methodology	Preferences in the contractor's engagement	Engagement of skilled contractors	3
	Pre-qualification of contractors and sub-contractors	Special requirements for the contractor in the tendering process	1
	Engagement of sustainability consultant		1
	Engagement of an independent commissioning agent		1
	Engagement of sustainability consultant in the selection of other team members		1
Timeliness of project activities	Early engagement of project team		5
	Early engagement of sustainability consultant		2
	Early engagement of design consultants		2
Complexity	Complexity in project development		10

Note: A detailed description of the themes related to differentiating conditions is provided in the following sections.

within all GBs and not at all in non-GB projects. In fact, these conditions are relatively stronger characteristics of GB project development; however, they may also occur in non-GB projects. Even the GB project development striving for low levels of sustainability may lag in some of these attributes as an interview participant (UK-M-5) highlighted,

> If the Green Building projects are executed in an integrated manner, then indeed they have a different delivery process from the traditional projects.

However, sometimes projects tagged as Green Building projects are executed in the same way as traditional projects. These are the kind of projects striving for low to medium level of sustainability performance.

1.2.1 Defining project goals

An important difference of GBs from non-GB projects is in terms of the project development goals as a UAE-based sustainability manager (AE-M-3) highlighted, 'unlike a traditional building, a Green Building project is required to be more efficient in terms of water and energy use. The focus on these different aspects makes a Green Building different from a traditional building'.

The difference between GBs and non-GBs begins to appear from the project initiation stage as the project brief is being developed. To ensure the sustainable development of a GB project, the project brief needs to incorporate green measures as an Australia-based design consultant (AU-M-2) highlighted, 'the difference would be in terms of the project brief which for a Green Building will need to have some description for the sustainability required in the building'. To define project aspirations, GBs also engage end-users in the project development as incorporating their expectations and needs significantly contributes towards the operational performance of these buildings.

These findings are supported by Hwang, Shan, and Supa'at (2017) according to whom GBs must accomplish additional goals in comparison with traditional buildings, namely energy, land, water, and material savings, as well as environmental friendliness.

1.2.2 Design methodology

A sharp contrast between the development of GBs and non-GB projects occurs at the stage of project design development. In traditional building projects, the architect develops the shape and decides the building materials. Afterwards, the building services engineers are involved who design the MEP services for the building. In contrast, a GB has different expectations and deliverables (for example, higher energy efficiency and environmental sustainability), and consequently, a GB differs from a traditional non-GB in terms of the design outputs as well as the design process. According to a UK-based sustainability consultant (UK-F-6), 'the development process of Green Building projects is different from traditional counterparts as in the Green Building projects additional design considerations and design constraints are added for the design consultants'.

An energy-efficient GB involves a rigorous design thinking to reduce the dependence on fossil fuels and overall energy use. Along with the typical functional planning required in a building, in GBs it is necessary to consider design aspects such as surroundings, orientation, and glazing.

Some of the prominent design development aspects differentiating GBs from non-GBs, outlined by interview participants, include the following:

- **The intensity of design activity:** Compared to a traditional building, in a GB the design activity is more intense, especially when a GB certification is being targeted. A high level of social sustainability on building projects requires excessive upfront consultation with the design team, which can also result in higher costs and longer delivery times. According to a UK-based design consultant (UK-M-6),

  > When the building is operated, it should be in a way that is genuinely low impact. A consultant needs a far bigger fee to do that work because it actually involves a lot more engagement and effort.

- Integration in design: GBs require more integrated design approaches. According to a UAE-based sustainability consultant (AE-M-1), 'Green Building projects dominantly have an integrated design process where a Green consultant coordinates among different design disciplines to ensure that all sustainable design aspects are implemented and coordinated within the design process'. In a GB, the architect develops an architectural design which is then used by MEP consultants to conduct preliminary modelling. This modelling provides an estimate of building efficiency upon which the architectural model is revised. The revised model is then provided to the structural and MEP consultants to develop the requisite drawings and models. Hence, a high degree of design integration is inherently required in GBs.
- Holistic and systems approach: GBs require a holistic and systems approach in design. These approaches are about seeing how building systems work together instead of being considered individually. According to an Australia-based commercial specification consultant (AU-M-21),

  > The difference of a 6-Star Green Star and a non-certified building will come, where the different design aspects are considered holistically or not. Buildings which are going for high-level green certifications look for a holistic approach, the integration of the building within the precinct, and the integration of occupants within the building. It all needs to be considered as a whole. A standard building might be looking at cabinetry, finishes, lobby, and main services related to works, but a Green Building will be looking at how the systems work together, in a systems approach. This is where the difference will lie. It is not something which is visual.

- Creativity in design: Projects opting for higher sustainability performance than minimum requirements need to follow a creative design approach. This is because owing to the sustainability requirements,

GBs face increased constraints and challenges which can only be mitigated by 'out of the box' approaches.

• **Material and technology selection:** The selection, sourcing, and procurement of building materials and technology can also set GBs apart from non-GB projects. Material supply and availability are more critical in GB construction projects compared to traditional building construction projects (Hwang & Leong, 2013; Hwang, Shan, Phua, & Chi, 2017). Usually, in traditional projects, all the procurement is locally performed since all the required materials are locally available. However, in the case of GBs, if sustainable materials are not locally available, they may need to be imported to fulfil sustainability requirements. GBs typically involve a search for materials and technologies with higher efficiencies, while traditional buildings usually depend on standard technology. Owing to the special material requirements in GBs, a high level of diligence is to be practised in material procurement. According to an Australia-based interview participant working as a design and BIM manager in a contractor organisation (AU-M-5),

> In case a project has sustainable objectives, a more stringent process is to be managed to result in project outcomes. This can particularly be a case when selecting the building materials and considering the embodied energy in material manufacturing.

The above-mentioned conditions differentiate GBs from non-GBs in terms of the design process. Some of the most prominent differentiating conditions are, however, in terms of design deliverables.

• **Life cycle perspective in design:** Unlike the case of traditional projects, which often have a short-term focus, a life-cycle perspective drives the process of GB project development. New materials in the market with lower environmental impacts and longer service lives are the favoured choices in GB development. The thought process involved in GB design may even consider the material reuse upon dismantling a building in future.

• **Designing for context, energy conservation, and end-users:** During the design development of GBs, social and environmental considerations in the design are required, which are otherwise ignored. For instance, the use of on-site energy generation is more prevalent in GB projects as compared to non-GBs. GB developments have a focus towards integrating the building within the surroundings and have emphasis on integrating users within the building.

1.2.3 Team procurement methodology

Unlike traditional buildings, GBs may involve experts such as sustainability consultants and independent commissioning agents to fulfil sustainability requirements. In case a sustainability consultant is hired before the architect and other consultants, s/he may help the client develop pre-qualification criteria for engaging other team members. However, in case the project team is already shortlisted, the sustainability consultant develops terms of reference, and the final selection of the project team has to fulfil these terms.

Compared to traditional projects, the expertise of key team members in GBs is very important. For instance, in such projects, it is very important to have the involvement of skilled contractors who can support sustainable development by reducing waste on-site, help in the sourcing of green construction materials, and provide their support in the documentation required for third-party GB certifications. GBs also have some special requirements that need to be fulfilled by contractors during tendering.

1.2.4 Timeliness of project activities

One of the major differences in the procurement of a GB from a traditional building is that a GB needs the involvement of team members early in project development as a Pakistan-based design and sustainability consultant (PK-M-1) highlighted, 'the major difference of the Sustainable project procurement from traditional project procurement is that the Sustainable projects involve different project trades early in the project'. A sustainability consultant needs to be involved in the early stages of GB development to ensure that the process of such a project is put in place by well-experienced team members and at an appropriate time. In addition to sustainability consultants, design consultants also need to be engaged in the early design stages of a GB project.

1.2.5 Team collaboration

The liaison among project team members and the way different team members interact becomes much more important in GBs as compared to traditional buildings. A very important aspect is the communication and the contribution of the whole project team as a UK-based design consultant (UK-F-2) highlighted,

> The most important difference [among GB and non-GB projects] is the communication and the contribution of the whole project team. In a traditional building, all the different team members are not much connected and work apart, while in a Green Building all the team members are well connected and approach the project development in a combined way.

Unlike traditional buildings, GBs face an increased requirement to be well integrated from the early design stages particularly in terms of the liaison between the mechanical engineer, structural engineer, and the architect. In GB projects, a sustainability consultant has to coordinate across the whole project team and s/he is to ensure that all sustainable design aspects are implemented within the design process across disciplines.

Integration and collaboration: The application of sustainability principles in building construction projects adds to the complexity of these projects, and thus a greater integration and collaboration among project team members is required as compared to traditional projects (Riley, Magent, & Horman, 2004). According to Rohracher (2001), contrarily to the case of traditional buildings, the successful development of highly efficient GBs considerably depends on a closer interaction of users, professionals, and suppliers. This is because compared to traditional buildings, in GBs, the variety of building components have much more complex and stronger mutual dependencies in functional terms. This consequently makes the GBs act like machines in certain ways.

1.2.6 Flow of project information

Joint meetings take place during the design development process of a GB in which team members try to solve each other's issues and resolve clashes in building services. Upon understanding the project requirements of a GB, the MEP consultants coordinate with the architect and iterative cycles of design activity are conducted before the design is finalised. Since non-GB projects may not require design optimisation for energy efficiency, the flow of project information in non-GB projects may not be as critical as for GB projects.

1.2.7 Documentation

GB projects as compared to non-GBs have more extensive documentation activity, often because of the third-party GB certifications. The documentation process benefits a project in many ways. For instance, information can be passed on to the facility management team. In case the facility management team is changed, the building record can be transferred to the substitute team, ensuring that the project information is preserved, and the operational performance of the project meets the aspirations.

While reflecting on Green Star, a sustainability manager (AU-F-1) said,

Achieving Green Star rating is a documentation-based process and people need to demonstrate the sustainable design and operation of buildings. Many times, things are not documented, and they are just inside people's heads, so an effort is required to document what actually takes place in a project. This is all good practice, but it may not be done in case a project is not pursuing Green Star rating.

1.2.8 Inspection, monitoring, and control

During the stages of project inspection and monitoring, the conditions which differentiate GBs from non-GBs are the commissioning and tuning requirements, and the needs of excessive metering.

- **Commissioning and tuning requirements:** GBs as compared to conventional building projects have an increased focus on commissioning and tuning. Commissioning and tuning ensure the optimum performance of building systems in the project's operational stage. Commissioning is promoted in GBs through the requirements of GB certification systems as an Australia-based sustainability consultant (AU-F-5) indicated, 'the sign off by an independent commissioning agent is something that was introduced by Green Star [in Australia]. Commissioning previously was focused on HVAC systems, but now it is being used for other systems as well'.
- **Monitoring and controlling operational performance:** GBs are different from non-GBs since they are focused towards monitoring and controlling the operational performance, and for this purpose, they incorporate excessive metering. Such metering enables quick feedback for the building users regarding the use of energy and water. Metering is emphasised so much in GBs that in the GB certification systems some credit points are dedicated to metering.

1.2.9 Budget and delivery time requirements

GBs as compared to their traditional counterparts have special considerations towards socio-economic and environmental aspects of the development. These considerations which distinguish GBs from non-GBs require additional time and budget. However, time and budget are not only the function of project sustainability but also depend on other conditions such as the timeliness of project activities.

1.2.10 Green Buildings as complex projects

The increased complexity in GB projects differentiates them from their traditional counterparts. Compared to traditional buildings, GBs encounter more complex problems during the construction process, and project managers are faced with greater challenges (Sang et al., 2018). The complexity of project development is itself attributable to mutually contradicting project deliverables, timeliness of project activities, and project design approach.

GB projects are more complex than non-GBs as they involve a reasonably higher number of deliverables. A UK-based design consultant (UK-F-5) highlighted, 'the level of green certification aspired for a GB project also matters as for a high level of certification, the requirements in terms of the delivery process are quite different. For instance, engagement of the project team, specialised MEP consultancy, etc.'.

The deliverables and constraints (for example, cost and environmental performance) in GBs may be mutually contradictory depending on certain conditions. It is challenging to optimise mutually conflicting deliverables for achieving the desired project performance. Hence, the increase in GB development complexity is attributable to a higher number of design constraints and deliverables. As a UK-based contractor (UK-M-7) said,

> The delivery of Green Building projects is complex compared to their traditional counterparts because more variables [liveability-related consideration, environment-related factors] need to be considered. When you have to build a Green Building, you have to consider many aspects other than just technical compliance. When you add more items into deliverables, it always makes it more complex.

An example of the complexity in GBs is the often contradicting relationship between environmental and economic sustainability. Since environment-friendly building components often have higher capital costs, they associate negatively with economic sustainability. The complexity resulting from such contradictory relationships among project deliverables renders many design and development alternatives unfeasible and therefore requires increased effort in developing project proposals which satisfy a number of deliverables.

Complexity in keeping a balanced approach

According to an Australia-based design consultant (AU-M-4),

> In case of a commercial project, if you increase the wall thickness two folds to get better insulation, it would mean that the leasable floor area drops. So, it is paramount to keep things in balance, otherwise, it may affect the project's feasibility. In designing a project, we are trying to find the golden mean.

The complexity in GB development can also be a function of the project stage at which sustainability is considered in a project. If it is a part of the project from the onset, there is a high level of possible optimisation and the number of constraints for incorporating green aspects are far less as compared to a situation where a project is partially or fully developed. For instance, if the base structure of a building has already been developed without considering sustainability, then the opportunity of appropriate building orientation and massing to reduce the operational energy use is already missed. Sustainability considerations at this stage may only be limited to the selection and sizing of the building systems which cannot offset the capital cost and energy savings, otherwise possible.

Sustainable performance of a GB may not always equate to technical complexity of the building. While the delivery and development of a GB project may be complex, the final output, that is the GB itself, maybe fairly simple and straightforward in terms of its installed systems, technology, and operations. Talking about this issue, a UK-based environmental designer (UK-F-4) said,

> In principle, Green Buildings need not be complex. As realised from many research studies and case studies, buildings with good environmental performance can be achieved by simple techniques such as a better layout, thoughtful placement of windows, local material sourcing and thermal properties of the building structure. For compliance, LEED and BREEAM rating tools do not require highly complex systems.

Although a GB does not need to be a technically complex end-product, owing to its mutually contradicting objectives (that is, realising environmental sustainability at low costs), the process of its development continues to be complex.

Complexity in passive designs

Hawken, Lovins, and Lovins (2013, p. 98) found that in terms of indoor environment quality and comfort, the systems used in some advanced buildings have low capital costs and surprisingly low energy use and noise accredited to the highly efficient fans and low-friction ducts they use. However, 'most innovative buildings' use computational fluid dynamics and do not even require fans as those buildings are designed to passively and noiselessly move the air through them.

Complexity depending on building typology

According to an Australia-based design consultant (AU-M-4),

> In commercial office buildings, the fit-outs are not necessarily as complex and diverse as compared to retail developments. For instance, in Chadstone there are 600 shops which means that it will have 600 different fit-outs. However, in a 20-storey office building you are probably dealing with 20 different tenants and therefore the variety of fit-outs will be limited. Hence, the complexity of managing the ultimate overall project is much higher in retail.

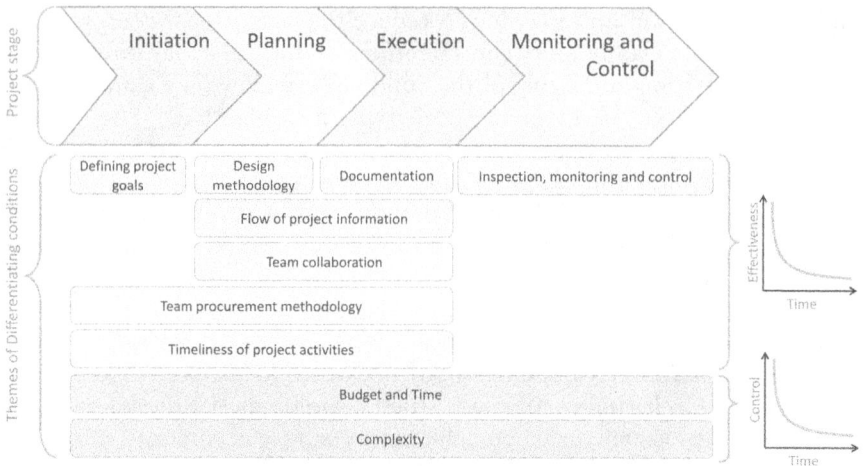

Figure 1.1 Differentiating conditions occurring in stages of the project lifecycle.

1.3 Occurrence of differentiating conditions during project stages

Plotting the occurrence of differentiating conditions during the project development (see Figure 1.1) indicates that in the initiation stage conditions regarding project goals prevail; in planning stage, design-related conditions prevail; during execution stage, procurement-related decisions are made; and upon project completion, inspection, monitoring, and control of operational performance become a differentiating attribute.

As shown in Figure 1.1, the project team's control over project 'complexity' as well as the 'budget and time' reduces as a project progresses from initiation to monitoring stage. Conditions such as 'Defining project goals', 'Design methodology', 'Documentation', 'Flow of project information', 'Inspection, monitoring and control', 'Team collaboration', 'Team procurement methodology', and 'Timeliness of project activities' affect the development of a GB project. However, their effectiveness in GB project development reduces as a project moves from its initiation to monitoring stage.

1.4 Associations among differentiating conditions

Among the differentiating conditions shown in Figure 1.1, some core conditions also lead to other conditions. For a better understanding of these core differentiating conditions, it is necessary to identify the associations among differentiating conditions. For instance, high budget and delivery time requirement of GBs is a differentiating condition. However, this condition is itself an effect of some other conditions such as design

Complexity
Team procurement methodology
(1, 4)
Design methodology
(5, 1)

(6, 1)

Defining project goals
(0, 5)
Budget and Time
(7, 1)
Flow of project information
(1, 3)

Timeliness of project activities
(2, 4)
Team collaboration
(2, 3)
Documentation
(1, 0)

Inspection, monitoring and control
(0, 2)

Legend
Themes representing Differentiating conditions
Low   Frequency   High
(Number of interview and survey participants)
(x, y)
Number of In-bound relationships
Number of Out-bound relationships

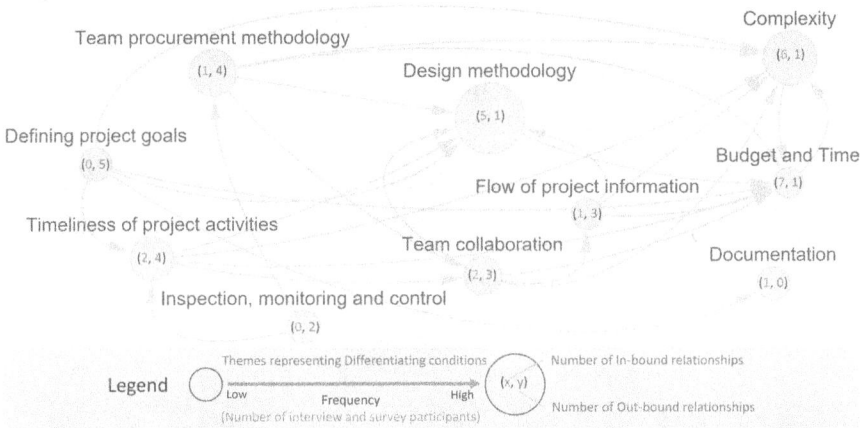

Figure 1.2 Associations among differentiating conditions identified from interview findings.

methodology. Interview findings have identified the associations among differentiating conditions as shown in Figure 1.2.

As the network (Figure 1.2) shows, 'timeliness of project activities', 'team procurement methodology', and 'defining project goals' are associated with 'design methodology', 'complexity', and 'budget and time'. Although 'Complexity in project development' is highly mentioned in interviews ($n = 10$), the associations suggest that complexity is determined by other aspects. The 'timeliness of project activities' and 'defining project goals' may affect 'complexity', which may eventually affect 'budget and time' as a UAE-based sustainability consultant (AE-M-1) indicated,

> Sustainability in such projects should ideally be considered during early design stages and such a practice can help avoid the cost overruns and time delays in Green Building projects. In the case of early incorporation of sustainable aspects, there would be no technical complexity involved in Green Building projects. The project client should formulate the project brief and requirements such that instead of considering sustainability as an add-on project feature, it should be considered a part of the project to reduce the technical complexity.

According to a UAE-based design consultant (AE-F-1),

> If a client decides to certify a building using green building standard and communicates it to the design team at a very early stage, it is easier and not complex to ensure sustainability implementation. However, in a lot of cases, this is not communicated to the team until it is too late. The design team then has to revise the design and amend it, increasing the complexity and cost.

The key associations among differentiating conditions (Figure 1.2) suggest that conditions related to 'complexity' and 'design methodology', as well as 'budget and time', are affected by conditions occurring in other themes such as 'team procurement methodology', 'timeliness of project activities', and 'defining project goals'. Procurement methodology, timeliness of activities, and defining goals can therefore be considered as the core aspects differentiating GBs from non-GBs. As shown in Figure 1.1, the effectiveness of 'timeliness of project activities', 'defining project goals', and 'team procurement methodology' is during the early project stages. 'Complexity' and 'Budget and Time' extend throughout the project development stages; however, the control of the project team over these reduces as the project moves to later stages.

'Procurement methodology', 'timeliness of activities', and 'defining project goals' are the core aspects differentiating GBs from non-GBs.

1.5 Overlap of success conditions and differentiating conditions

Success conditions are the GB-project-related attributes contributing towards its success. These conditions are discussed in detail in Part II of this book. Upon matching the differentiating conditions with success conditions, it is found that as many as 22 (65%) differentiating conditions also occur as success conditions for GBs. The following attributes distinguishing GBs from non-GB projects have much significance since these differentiating conditions are also success enabling factors:

- Setting of a detailed sustainability charter or brief
- Level of complexity in project design
- Life-cycle-based project development approach
- Use of energy-oriented design approach
- Use of integrated design approach
- Monitoring and controlling operational performance of building
- Execution of commissioning and fine-tuning
- Preferences in contractor's engagement
- Design consideration towards future building usage
- Use of effective communication tools and strategies
- End-users' involvement in defining project aspirations
- Use of context-oriented design approach
- Use of holistic design approach
- Use of innovative design approach

- Market survey to identify successful building systems
- Project team collaboration
- Liaison between sustainability consultant and project team
- Pre-qualification of contractors and sub-contractors
- Involvement of sustainability consultant in contractor's selection
- Early engagement of project team
- Early engagement of design consultants
- Early engagement of sustainability consultant

1.6 Discussion

Among differentiating conditions, some explicitly account for sustainability in a project, namely 'setting of a detailed sustainability charter or brief', 'liaison between sustainability consultant and project team', 'engagement of sustainability consultant', 'engagement of sustainability consultant in selection of other team members', 'use of energy-oriented design approach', 'life-cycle based project development approach', and 'early engagement of sustainability consultant'. Apart from these conditions and the conditions related to budget, time, and complexity, all the other conditions are about effective project development practices. In the light of the above, it can be stated that GBs are different from non-GBs mainly because of the approaches necessary for their development.

This also implies that compared to non-GBs, for a GB project the use of effective project development approaches is very critical. Since a GB is more likely to follow effective project development approaches, it can mean that a GB project may have a higher quality performance compared to a non-GB project. While these conditions can result in a good performance in non-GB projects, their significance is much higher when it comes to GB projects. These differentiating conditions appear as indispensable for the development of GB projects.

> For a GB project, the use of effective project development approaches is very critical. Since a GB is more likely to follow effective project development approaches, it can mean that a GB project may have a higher quality performance compared to a non-GB project.

A recurrent theme in the interviews was a sense among interviewees that the sustainability requirements which differentiate GBs from traditional buildings are not a sufficient reason for the construction industry to rationalise the need for radically different project development approaches for GBs. Based on this, it can be assumed that the construction industry will continue to use the existing project development approaches for GBs. For instance, similar to the non-GB projects, GBs are also developed using popular project delivery methods (PDMs) in the construction industry (for example, Design-Build, Design-Bid-Build, and Construction Management at Risk). As a Singapore-based commissioning service provider (SN-M-2) highlighted, 'the contractual approach to be used in the project is not often decided on whether the building is to be green or not, it is decided based on project size'. According to another interview participant (AU-M-5), 'any delivery method can be used for Green Building projects as long as the ecologically sustainable development requirements are contractually binding in both the head contract and all subcontracts'. There are no reported cases in reviewed literature and interview data which state that a GB project was developed based on a PDM especially designed for GBs. On the contrary, many studies have been conducted to explore the effectiveness of already existing PDMs for GB projects.

From the interview findings, it is realised that the construction industry professionals are more inclined to think of GBs as construction projects with sustainability requirements rather than thinking of them as sustainable development projects. Thinking of GBs as construction projects helps in their rapid adoption since the readily available project development approaches from the construction industry will drive the process of their development. However, considering GBs as construction projects also means that GBs inherit the constraints and challenges of a typical construction project such as the loss of productivity and the wastefulness of resources. On the contrary, in case construction industry professionals think of GBs as the outcomes of sustainable development instead of construction, the paradigm shift can open many avenues of thought which can help achieve higher levels of sustainability potentials in GBs. As Albert Einstein said, 'We cannot solve our problems with the same thinking we used when we created them'.

In this chapter, the conditions differentiating GBs from non-GBs are explained. These two types of projects occur at the extreme ends of the spectrum, and there can be many variations of these projects in which the differentiating conditions may not be much prominent. For instance, a non-GB project may be developed using a highly integrated design, while a GB may not.

> In case construction industry professionals think of GBs as the outcomes of sustainable development instead of construction, the paradigm shift can open many avenues of thought which can help achieve higher levels of sustainability potentials in GBs.

1.7 Summary

In this chapter, conditions differentiating GBs from non-GBs are identified based on the opinions of 25 GB experts. The identified conditions are grouped in ten themes, which are 'Budget and Time' (number of participants = 1), 'Defining project goals' ($n = 2$), 'Design methodology' ($n = 16$), 'Documentation' ($n = 2$), 'Flow of project information' ($n = 1$), 'Inspection, monitoring and control' ($n = 2$), 'Team collaboration' ($n = 4$), 'Team procurement methodology' ($n = 7$), 'Timeliness of project activities' ($n = 6$), and 'Complexity' ($n = 10$). The highest number of differentiating conditions are related to 'design methodology' (that is, 11) and 'team procurement methodology' (that is, 5). The overall analysis of differentiating conditions suggests that GBs are primarily different from non-GBs because of the effective project development approaches they require.

The associations among differentiating conditions are analysed, and these show that the conditions related to 'procurement methodology', 'timeliness of activities', and 'defining project goals' are the core aspects differentiating GBs from non-GBs. Conditions related to 'complexity' and 'design methodology', as well as 'budget and time', are affected by 'team procurement methodology', 'timeliness of project activities', and 'defining project goals'. The overlap among success conditions and differentiating conditions is also analysed to interpret the significance of differentiating conditions.

This chapter has addressed the research question, that is *How do sustainability requirements set Green Buildings apart from non-Green Buildings?* But *What are Success conditions in Green Building projects and how do they interrelate?* Part II will answer this question.

References

Hawken, P., Lovins, A. B., & Lovins, L. H. (2013). *Natural capitalism: The next industrial revolution*. London: Routledge.

Hwang, B. G., & Leong, L. P. (2013). Comparison of schedule delay and causal factors between traditional and green construction projects. *Technological and Economic Development of Economy, 19*(2), 310–330. doi:10.3846/20294913.2013.798596

Hwang, B. G., Shan, M., Phua, H., & Chi, S. (2017). An exploratory analysis of risks in green residential building construction projects: The case of Singapore. *Sustainability, 9*(7), 1116. doi:10.3390/su9071116

Hwang, B. G., Shan, M., & Supa'at, N. N. B. (2017). Green commercial building projects in Singapore: Critical risk factors and mitigation measures. *Sustainable Cities and Society, 30*, 237–247. doi:10.1016/j.scs.2017.01.020

Riley, D. R., Magent, C. S., & Horman, M. J. (2004, May 2–7). Sustainable metrics: A design process model for high performance buildings. Paper presented at the CIB World Building Congress, CIB, Toronto.

Rohracher, H. (2001). Managing the technological transition to sustainable construction of buildings: A socio-technical perspective. *Technology Analysis & Strategic Management*, *13*(1), 137–150. doi:10.1080/09537320120040491

Sang, P., Liu, J., Zhang, L., Zheng, L., Yao, H., & Wang, Y. (2018). Effects of project manager competency on green construction performance: The Chinese context. *Sustainability*, *10*(10), 3406. doi:10.3390/su10103406

Part II

Green Building project success

This part aims to address the question, that is *What are success conditions in Green Building projects and how do they interrelate?*

Part II has four chapters, beginning with Chapter 2 which provides a detailed account of social success conditions in GB projects. This is followed by Chapter 3 which is about the GB success conditions mainly possessing a technical dimension. Subsequently, in Chapter 4 the success conditions introduced earlier are interpreted both in terms of theory and practice. This provides an understanding of the value associated with GB success conditions for both the theory and practice of the built environment. Limitations of the study and recommendations for future studies are also provided. Lastly, Chapter 5 provides a detailed account of the stakeholders affecting and affected by GB success conditions. Suggestions for key project stakeholders to ensure successful GB project development are also provided.

A brief overview of the methodology and the highlights of the GB success-related findings are provided in the following sections.

PII.I Overview of research design

Following an exploratory research design, the inquiry in this part is based on the in-depth analysis of qualitative data obtained from semi-structured interviews. The thematic analysis conducted on interview data helped identify the success conditions for GB projects. Using the in-depth understanding of GB success conditions provided in interviews, the importance of different conditions is also noted.

The 75 interviews participants responding to the success conditions *theme* were based in Australia ($n = 32$; 42%), Hong Kong ($n = 8$; 11%), Pakistan ($n = 2$; 3%), Singapore ($n = 14$; 19%), the UAE ($n = 6$; 8%), and the UK ($n = 13$; 17%). The interview participants are highly experienced building professionals as most of them ($n = 49$; 68%) have a minimum of ten years of experience in developing GB projects (see Figure PII.1). The participants are mainly design/architecture consultants ($n = 19$; 26%), and sustainability consultants ($n = 22$; 29%). The interview participants also

DOI: 10.1201/9781003322740-3

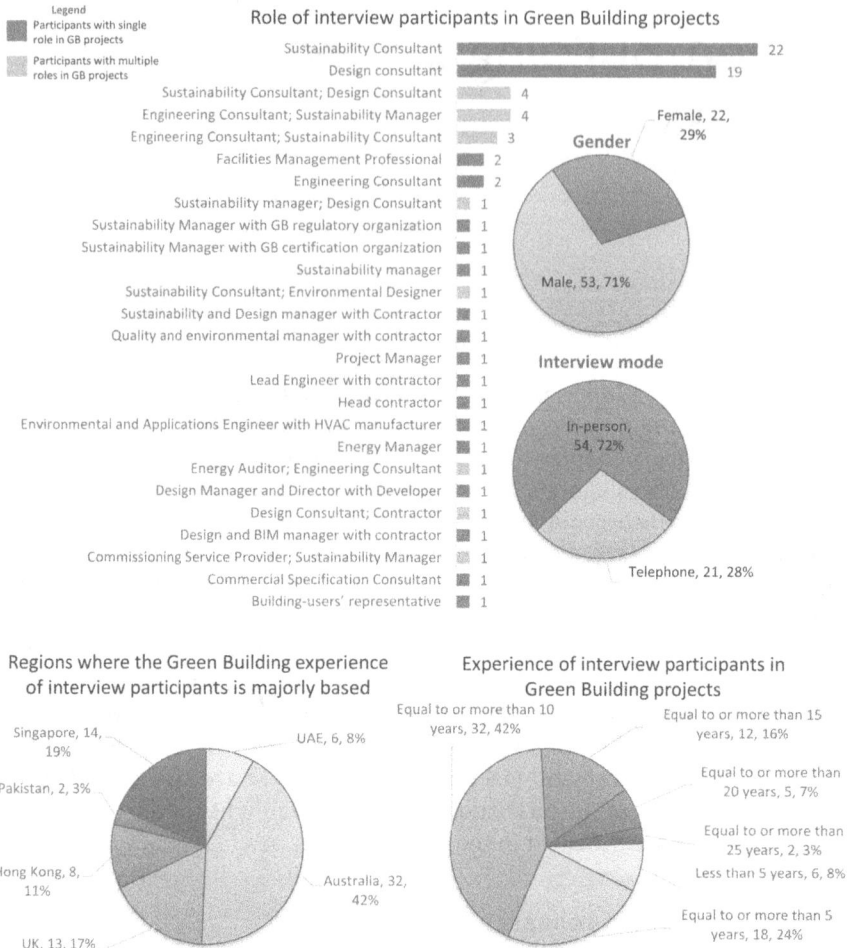

Figure PII.I   Role, experience, and regional belonging of interview participants discussing success conditions.

had the roles of clients/developers, contractors, project managers (PMs), and facility managers in GB projects. Up to 32 (44%) of these participants had experience in acting in multiple roles during the development of GB projects. A detailed overview of data collection from interviews is provided in Annex A (Section 6.2).

Chapters 2 and 3 present the interview findings obtained from three viewpoints used in semi-structured interviews. The viewpoint of *Findings from participants' overall experience* was intended to highlight success conditions for GBs, the interview participants had experienced in their overall professional careers of developing such projects. *Findings from successful/*

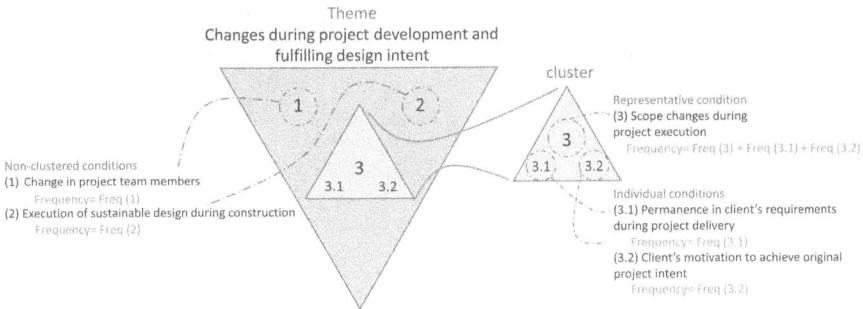

Theme
Changes during project development and
fulfilling design intent

cluster

Representative condition
(3) Scope changes during
project execution
Frequency= Freq (3) + Freq (3.1) + Freq (3.2)

Non-clustered conditions
(1) Change in project team members
Frequency= Freq (1)
(2) Execution of sustainable design during construction
Frequency= Freq (2)

Individual conditions
(3.1) Permanence in client's requirements
during project delivery
Frequency= Freq (3.1)
(3.2) Client's motivation to achieve original
project intent
Frequency= Freq (3.2)

Figure PII.2 Occurrence of success conditions and sub-conditions within clusters and themes.

failed projects highlighted the success and failure conditions of projects occurring in the professional careers of participants. *Suggestions for clients* provided success conditions in control of the client developing the project. A detailed explanation regarding these viewpoints is provided in research design (Annex A – Section 6.3).

The interview data were analysed using thematic analysis, which helped the identification of success conditions and their associations. A hierarchical structure is developed for rigorous, transparent, and detailed analysis and reporting of conditions shown in Figure PII.2. In this hierarchy, a *subcondition* is part of a success condition. A success condition and its subconditions collectively make a *cluster*. Multiple success conditions constitute a *theme* which is a classification introduced to collectively explain similar conditions. For instance, 'Changes during project development and fulfilling design intent' is a *theme* comprised of three success conditions. One of the three success conditions is 'Scope changes during project execution', which represents a *cluster* of two sub-conditions which are 'Permanence in client's requirements during project delivery' and 'Client's motivation to achieve original project intent'.

The detailed analysis of interview findings identified 155 conditions and sub-conditions associated with GB project success. Overall, there are 25 clusters which include 1–15 sub-conditions. While 25 success conditions and 82 sub-conditions occur in clusters, the rest of 48 success conditions do not represent sub-conditions and therefore do not occur in clusters. The overall 73 success conditions and their corresponding 82 sub-conditions are allocated across 20 broad *themes* which emerged from the analysis.

The number of times a success condition is highlighted by interview participants, that is, 'frequency', is an important parameter for analysis. As shown in Figure PII.2, the frequency of a success condition in a *cluster* is the cumulative frequency of the whole *cluster* (that is, success condition and sub-condition/s). For instance, 'Scope changes during project execution' is the success condition of a *cluster* with two sub-conditions, which are 'Permanence in client's requirements during project delivery' and 'Client's motivation to achieve original project intent'. Frequency of 'Scope changes during project execution' is, therefore, the cumulative frequency of the overall *cluster* (that is, 9).

PII.2 Overview of success conditions and their interrelationships

Before providing in-depth analysis of individual success conditions, this section provides a brief introduction to the identified success conditions occurring within the broader *themes*. In this section, the reasons behind the importance of success conditions are also discussed based on the inter-relationships between success conditions.

Key trends emerge by observing the identified success conditions at the level of broader *themes*, as shown in Figure PII.3. From among the 20 *themes*, nine are directly related to project stakeholders and most of these are related to the project team. These *themes* are about team characteristics, collaboration, commitment, mindset, authorities, responsibilities, and procurement methodology. Some of the stakeholder-related *themes* include the client's characteristics; education of project team, client, and end-users; and the cooperation and interest of external stakeholders.

The remaining 11 *themes* are of technical nature and are indirectly related to project stakeholders. These technical conditions focus mostly on using effective and rigorous methodologies at different project stages, for instance, project goal definition, planning, design development, inspection, monitoring, and control. Some of the conditions are not specific to a

Figure PII.3 Number of participants identifying conditions within broader themes.

project stage and instead apply to the entire process of project development, for instance, clarity in the development process, the flow of project information, completeness and rigour of deliverables, and managing project changes. Some of the technical conditions are also the project constraints which are beyond the control of the project team and can also be beyond the client's control. Together, the nine *themes* directly related to project stakeholders and the 11 *themes* of technical nature represent a myriad of 155 socio-technical success conditions for GB projects.

The in-depth interview data collection investigating success conditions has determined 'why' the identified conditions are important for the success of GB projects. Upon conducting a detailed analysis of interview data, the reasons of importance for each condition are identified. Within this list of reasons, 80% (that is, 74) of the reasons are the factors which are already identified as success conditions. Hence, in the majority of cases, the reason for the importance of a success condition is that it results in or associates with other success conditions. For instance, the importance of 'clarity in process of developing project' (one success condition) is that it results in the 'alignment of team interest with project interest' (another success condition).

Success conditions and sub-conditions and their relevant themes are shown in Table PII.1. This shows the 73 success conditions and 82 subconditions identified in this study. It also indicates 93 success conditions and sub-conditions which are not only identified from expert interviews in this study but have also been studied in previous publications. Among the 155 success conditions and sub-conditions identified in this study, 62 success conditions and sub-conditions are unique (that is, have not been discussed in previous studies as GB success conditions/factors). A detailed overview of these conditions is provided in Chapters 2 and 3. The associations of success conditions and sub-conditions are collectively shown through a network diagram in Figure PII.4 in which the edges (that is, associations) between nodes (that is, conditions) tend to imply the reasons for their importance. In such a network, an edge from node A to B not only implies the association among the two conditions, it also means that node A (that is, condition A) is important since it is related to node B. These associations in case of each broader *theme* of success conditions are explained in detail in the individual *theme* sections (in Chapters 2 and 3). Each section in these chapters provides a brief explanation of success conditions regarding the 20 broader *themes*. Each section represents a *theme* and is presented according to the following structure: a figure depicts the conditions occurring in a *theme* and the reasons for the importance of those conditions; the key conditions occurring in a *theme* are mentioned; the associations of conditions with project stakeholders are mentioned; and the associations of conditions with the Transformation, Flow, and Valuegeneration (TFV) views of the TFV theory are explained.

Table PII.I Green Building success conditions

Theme	Sr. no.	Identified conditions and sub-conditions	Overall interviewees identifying	Occurrence in previous studies
Changes	I	Change in project team members	2	
during project	2	Execution of sustainable design during construction	5	×
development				
and fulfilling	3	Scope changes during project execution	9	×
design intent	3.I	Permanence in client's requirements during project delivery	2	×
(n = 14)				
	3.2	Client's motivation to achieve original project intent	3	
Clarity in	4	Clarity in process of developing project	9	×
project	4.I	Use of clearly defined and standardised approaches for GB development	5	×
development				
(n = 13)	5	Delegating clear responsibilities to project team	6	×
	5.I	Project team contractually required to deliver sustainable outcomes	2	×
	5.2	Provision of sustainability specifications and other related information in tender	4	
	5.3	Specificity of deliverables from design consultants	2	
Client's	6	Client's leadership in project	3	
characteristics	7	Consensus within client organisation	2	×
(n = 24)	8	Proficiency of project client	20	×
	8.I	Client understanding the need of sustainable outcomes	5	×
	8.2	Client's understanding of GB requirements	I2	×
	8.3	Client's understanding of sustainable building operation	2	
	8.4	Client's rational decision-making	I	×
	9	Structure and nature of client organisation	5	×
Completeness	I0	Completeness of project documentation for certification	2	
and rigour of				
deliverables	I I	Completeness and rigour of project design before execution	9	×
(n = 20)	I2	Setting of a detailed sustainability charter or brief	I0	
Constraints	I3	Access to sustainable building materials	2	×
(n = 9)	I4	Accessibility of project funding	6	×
	I5	Ease of logistics at project location	I	×
Cooperation	I6	Client's involvement in project development	5	×
and interest of				
stakeholders	I6.I	Client's facilitation of coordination among project team	I	
(n = 42)				
	I7	Client's motivation to achieve sustainable outcomes	32	×
	I7.I	Client's endorsement of sustainability brief	3	

(Continued)

Table PII.1 Green Building success conditions (*Continued*)

Theme	Sr. no.	Identified conditions and sub-conditions	Overall interviewees identifying	Occurrence in previous studies
	17.2	Drivers for client to achieve sustainable outcomes	12	×
	17.3	Investor's motivation to achieve sustainable outcomes	1	×
	18	Cooperative role of building control authorities	2	
	19	End-users' operation of building in sustainable ways	12	×
	19.1	End-users' motivation to achieve sustainable outcomes	1	×
	19.2	End-users' understanding of building operation	5	×
	19.3	Tenants contractually required to consider sustainability	2	
	20	Stakeholders' approval of project	4	×
Defining project goals (*n* = 31)	21	End-users' involvement in defining project aspirations	8	×
	21.1	FM team involvement in defining project aspirations	1	
	22	Project team involvement in defining project aspirations	3	
	23	Setting appropriate project targets	21	×
	23.1	Clarity in building's operational performance targets	1	
	23.2	Nature of GB certification aspired for project	2	
	23.3	Sustainability brief aligned with project budget	3	×
	23.4	Specificity of project requirements	12	×
	24	Stringency level of project sustainability requirements	4	×
	24.1	Stringency level of GB certification requirements	2	
Design methodology (*n* = 41)	25	Market survey to identify successful building systems	1	
	26	Rigour of project design development	41	×
	26.01	Adding adaptability and multiple layers of use in building design	1	×
	26.02	Use of reliable technology and solutions	2	×
	26.03	Design consideration towards future building usage	3	
	26.04	Level of complexity in project design	4	×
	26.05	Life-cycle-based project development approach	4	×
	26.06	Maintainability considered in building design	3	

(*Continued*)

Table PII.I Green Building success conditions (*Continued*)

Theme	Sr. no.	Identified conditions and sub-conditions	Overall interviewees identifying	Occurrence in previous studies
	26.07	Speculation in building design	7	
	26.08	Suitability of the project design for execution	3	x
	26.09	Use of a proactive design approach	5	
	26.1	Use of a balanced design approach	1	
	26.11	Use of a context-oriented design approach	3	
	26.12	Use of an energy-oriented design approach	12	x
	26.13	Use of a holistic design approach	9	x
	26.14	Use of an innovative design approach	6	
	26.15	Use of an integrated design approach	13	x
	27	Use of performance-based specifications	1	
Education (n = 18)	28	Educating client about sustainability in project	6	x
	29	Educating end-users and FM team about building operation	9	x
	30	Educating project team about GB development	8	
	30.1	Educating contractors about GB development	3	x
	30.2	Educating sub-contractors about GB development	2	x
Engraving sustainability in project development (n = 9)	31	Engraving sustainability in project development	8	
	32	Procurement of project site based on sustainability goals	3	x
	33	Use of environmental management systems in construction	1	
Flow of project information (n = 25)	34	Clarity in communication of project goals	7	
	35	Communication among project team	11	x
	35.1	Use of effective communication tools and strategies	5	x
	36	Project team's access to robust information	5	x
	36.1	Sharing of information related to project changes	1	x
	37	Smooth transition of project from inception to operation stage	5	
Inspection, monitoring, and control (n = 25)	38	Inspection of project upon construction	8	
	38.1	Execution of commissioning and fine-tuning	7	x
	39	Monitoring and controlling operational performance of building	5	x
	40	Monitoring of project development	12	x
	40.1	Project reporting	2	x
	41	Review of project design by sustainability consultant	2	
	42	Thoroughness of value engineering exercise	3	

(*Continued*)

Table PII.1 Green Building success conditions (*Continued*)

Theme	Sr. no.	Identified conditions and sub-conditions	Overall interviewees identifying	Occurrence in previous studies
Planning	43	Attention towards details	5	×
approach	44	Rigour of project planning	19	×
(n = 24)	44.1	Adequate budget allocation for project development	13	×
	44.2	Adequate time allocation for commissioning and testing	1	×
	44.3	Adequate time allocation for project development	11	×
	45	Rigour of risk management	4	×
Team	46	Control of project design by design and sustainability consultant	2	×
authorities,				
responsibilities,	47	Empowerment of sustainability consultant by client	3	
and				
contractual	48	Project team's involvement in decision-making	2	
relationships				
(n = 9)	49	Using appropriate project delivery method	3	×
	49.1	Contractual interrelationships between client and project team	2	×
Team	50	Like-mindedness of project team members	2	
characteristics	51	Proficiency of FM team	3	×
(n = 36)	52	Proficiency of project team	32	×
	52.1	Leadership qualities among project team members	2	×
	52.2	Proficiency of contractor	12	×
	52.3	Proficiency of design consultant	12	×
	52.4	Proficiency of MEP consultant	6	
	52.5	Proficiency of PM team	4	×
	52.6	Proficiency of sub-contractor	5	×
	52.7	Proficiency of sustainability consultant	5	×
	53	Project team's understanding of project goals and aspirations	3	×
	54	Size of design team	2	
Team	55	Conflicts among project team	1	
collaboration	56	Project team collaboration	29	×
(n = 30)	56.1	Liaison between design team and client	5	×
	56.2	Liaison between FM team and project development team	2	×
	56.3	Liaison between sustainability consultant and client	1	
	56.4	Liaison between sustainability consultant and contractor team	3	
	56.5	Liaison between sustainability consultant and design consultant	4	
	56.6	Liaison between sustainability consultant and project team	5	
	56.7	Willingness of project team to work together	4	

(*Continued*)

Table PII.1 Green Building success conditions (*Continued*)

Theme	Sr. no.	Identified conditions and sub-conditions	Overall interviewees identifying	Occurrence in previous studies
Team commitment to the project (n = 27)	57	Alignment of team interest with project interest	9	x
	57.1	Rewards for achieving performance targets	2	x
	58	Contractor's proactive role in project development	2	
	59	FM team motivation to achieve sustainable outcomes	1	
	60	Project team motivation to achieve sustainable outcomes	20	x
	60.1	Contractor team motivation to achieve sustainable outcomes	8	
	60.2	Design consultant's motivation to achieve sustainable outcomes	10	x
	60.3	MEP consultant's motivation to achieve sustainable outcomes	3	x
	60.4	PM team motivation to achieve sustainable outcomes	3	x
	60.5	Sustainability consultant's motivation to achieve sustainable outcomes	2	
Team mindset and priorities (n = 18)	61	Establishing and promoting synergies	2	x
	62	Focus of sustainability consultant on project goals	1	
	63	Open-mindedness and flexibility of project team	3	
	64	Priority of sustainability in project development	9	x
	65	Team working on project with innovative mindset	1	x
	66	Team working on project with value management mindset	5	x
Team procurement methodology (n = 13)	67	Involvement of sustainability consultant in contractor's selection	1	
	68	Preferences in project team selection	12	x
	68.1	Long-term engagement of sustainability consultant	2	x
	68.2	Preferences in consultant's engagement	5	
	68.3	Preferences in contractor's engagement	5	x
	68.4	Pre-qualification of contractors and sub-contractors	1	
	68.5	Requirement for contractor to engage a sustainability advisor	1	

(*Continued*)

Table PII.1 Green Building success conditions (*Continued*)

Theme	Sr. no.	Identified conditions and sub-conditions	Overall interviewees identifying	Occurrence in previous studies
Timeliness of project activities (n = 39)	**69**	Early engagement of project team	30	x
	69.1	Early engagement of commissioning professionals	3	x
	69.2	Early engagement of contractor	6	x
	69.3	Early engagement of design consultants	1	
	69.4	Early engagement of FM team	6	x
	69.5	Early engagement of sub-contractor	1	
	69.6	Early engagement of suppliers	2	
	69.7	Early engagement of sustainability consultant	22	
	70	Early introduction of project targets	22	
	70.1	Client's early decision-making regarding sustainability goals	4	
	70.2	Early incorporation of sustainability in project	17	x
	71	Timeliness of building approval	2	x
	72	Timeliness of feedback on sustainability documentation	1	
	73	Timely submission of GB certification documentation for review	2	x
		Number of interview participants	75	
		Number of conditions identified by interview participants	73	
		Number of sub-conditions identified by interview participants	82	

Note: Previous studies in which some of the similar success conditions are investigated: Aktas and Ozorhon (2015); Aktas, Ryan, Sweriduk, and Bilec (2012); Bakar, Abd Razak, Abdullah, Awang, and Perumal (2010); Banihashemi, Hosseini, Golizadeh, and Sankaran (2017); Bilec (2008); Bilec et al. (2009); Bond (2010); Carpenter (2005); Dahl, Horman, Pohlman, and Pulaski (2005); Doskočil and Lacko (2018); El Asmar, Hanna, and Loh (2013); Elforgani, Alnawawi, and Rahmat (2014); Enache-Pommer and Horman (2009); Gard (2004); Gultekin, Korkmaz, Riley, and Leicht (2013); Hanks (2015); Hwang and Ng (2013); Hwang, Zhao, and Tan (2015); Hwang, Zhu, and Ming (2016); Hwang, Zhu, and Tan (2017); Ihuah, Kakulu, and Eaton (2014); Kiani Mavi and Standing (2018); Korkmaz, Horman, Molenaar, and Gransberg (2010); Korkmaz, Riley, and Horman (2010); Korkmaz, Riley, and Horman (2011); Korkmaz, Swarup, and Riley (2011); Lam, Chan, Poon, Chau, and Chun (2010); Lei, (2005); Li, Chen, Chew, and Teo (2014); Li, Chen, Chew, Teo, and Ding (2011); Molenaar et al. (2009); Murtagh, Roberts, and Hind (2016); Olanipekun, Xia, Hon, and Darko (2018); Oyebanji, Liyanage, and Akintoye (2017); Pheng Low, Gao, and Lin Tay (2014); Rasekh and McCarthy (2016); Riley, Sanvido, Horman, McLaughlin, and Kerr (2005); Robichaud and Anantatmula (2010); Shen et al. (2017); Swarup, Korkmaz, and Riley (2011); Tang, Ng, and Skitmore (2019); Venkataraman and Cheng (2018); Wai, Yusof, Ismail, and Tey (2012); Wang, Wei, and Sun (2013); Xia, Skitmore, Wu, and Chen (2014); Zhang, Li, and Zhou (2017); Zhou and Smith (2013).

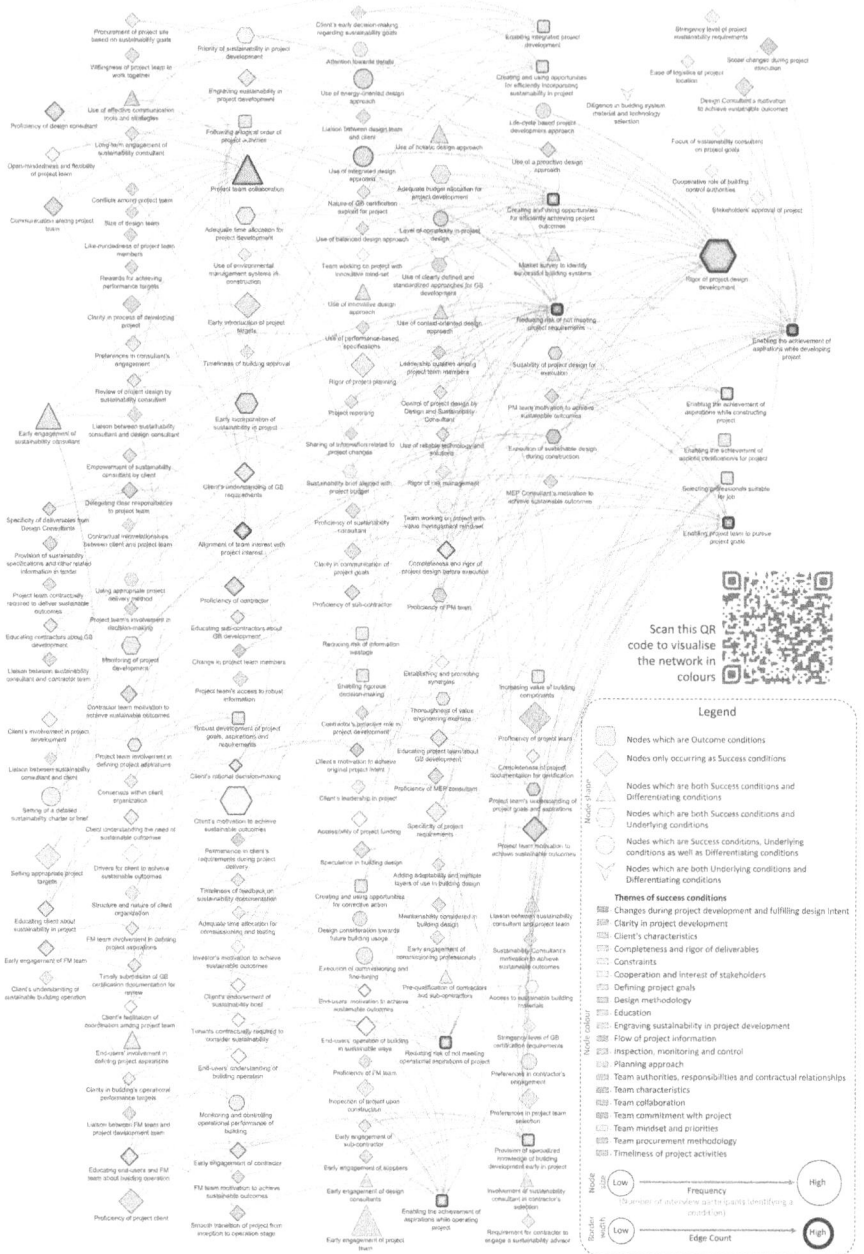

Figure PII.4 Network of Green Building success conditions.

References

Aktas, B., & Ozorhon, B. (2015). Green building certification process of existing buildings in developing countries: Cases from Turkey. *Journal of Management in Engineering, 31*(6). doi:10.1061/(ASCE)ME.1943-5479.0000358

Aktas, C. B., Ryan, K. C., Sweriduk, M. E., & Bilec, M. M. (2012). Critical success factors to limit constructability issues on a net-zero energy home. *Journal of Green Building, 7*(4), 100–115. doi:10.3992/jgb.7.4.100

Bakar, A. H. A., Abd Razak, A., Abdullah, S., Awang, A., & Perumal, V. (2010). Critical success factors for sustainable housing: A framework from the project management view. *Asian Journal of Management Research, 1*(1), 66–80. Retrieved from http://www.ipublishing.co.in/ajmr.html

Banihashemi, S., Hosseini, M. R., Golizadeh, H., & Sankaran, S. (2017). Critical success factors (CSFs) for integration of sustainability into construction project management practices in developing countries. *International Journal of Project Management, 35*(6), 1103–1119. doi:10.1016/j.ijproman.2017.01.014

Bilec, M. M. (2008). *Investigation of the relationship between green design and project delivery methods.* Retrieved from https://www.osti.gov/servlets/purl/937581

Bilec, M. M., Ries, R. J., Needy, K. L., Gokhan, M., Phelps, A. F., Enache-Pommer, E., … McGregor, E. (2009). Analysis of the design process of green children's hospitals: Focus on process modeling and lessons learned. *Journal of Green Building, 4*(1), 121–134. doi:10.3992/jgb.4.1.121

Bond, S. (2010). Lessons from the leaders of green designed commercial buildings in Australia. *Pacific Rim Property Research Journal, 16*(3), 314–338. doi:10.1080/14445921.2010.11104307

Carpenter, D. S. (2005). Effects of contract delivery method on the LEED (trademark) score of US navy military construction projects (Fiscal years 2004–2006) (Master dissertation). Oregon State University, Monterey, CA.

Dahl, P., Horman, M., Pohlman, T., & Pulaski, M. (2005). Evaluating design-build-operate-maintain delivery as a tool for sustainability. Paper presented at the Construction Research Congress 2005: Broadening Perspectives, San Diego, CA.

Doskočil, R., & Lacko, B. (2018). Risk management and knowledge management as critical success factors of sustainability projects. *Sustainability, 10*(5), 1438. doi:10.3390/su10051438

El Asmar, M., Hanna, A. S., & Loh, W.-Y. (2013). Quantifying performance for the integrated project delivery system as compared to established delivery systems. *Journal of Construction Engineering and Management, 139*(11), 04013012. doi:10.1061/(ASCE)CO.1943-7862.0000744

Elforgani, M. S. A., Alnawawi, A., & Rahmat, I. B. (2014). The association between client qualities and design team attributes of green building projects. *ARPN Journal of Engineering and Applied Sciences, 9*(2), 160–172. Retrieved from http://www.arpnjournals.com/jeas/

Enache-Pommer, E., & Horman, M. (2009). Key processes in the building delivery of green hospitals. Paper presented at the Construction Research Congress, Seattle, WA.

Gard, P. T. (2004). Fast and innovative delivery of high performance building: Design-build delivers with less owner liability. *Strategic Planning for Energy and the Environment, 23*(4), 7–22. doi:10.1080/10485230409509647

Gultekin, P., Korkmaz, S., Riley, D. R., & Leicht, R. M. (2013). Process indicators to track effectiveness of high-performance green building projects. *Journal of Construction Engineering and Management, 139*(12), A4013005. doi:10.1061/(ASCE)CO.1943-7862.0000771

Hanks, N. M. (2015). *Investigation into the effects of project delivery methods on LEED targets* (MSc thesis). University of San Francisco, San Francisco, CA. Retrieved from http://repository.usfca.edu/cgi/viewcontent.cgi?article=1152&context=capstone

Hwang, B. G., & Ng, W. J. (2013). Project management knowledge and skills for green construction: Overcoming challenges. *International Journal of Project Management, 31*(2), 272–284. doi:10.1016/j.ijproman.2012.05.004

Hwang, B. G., Zhao, X., & Tan, L. L. G. (2015). Green building projects: Schedule performance, influential factors and solutions. *Engineering, Construction and Architectural Management, 22*(3), 327–346. doi:10.1108/Ecam-07-2014-0095

Hwang, B. G., Zhu, L., & Ming, J. T. T. (2016). Factors affecting productivity in green building construction projects: The case of Singapore. *Journal of Management in Engineering, 33*(3), 04016052. doi:10.1061/(ASCE)ME.1943-5479.0000499

Hwang, B. G., Zhu, L., & Tan, J. S. H. (2017). Identifying critical success factors for green business parks: Case study of Singapore. *Journal of Management in Engineering, 33*(5). doi:10.1061/(ASCE)ME.1943-5479.0000536

Ihuah, P. W., Kakulu, I. I., & Eaton, D. (2014). A review of Critical Project Management Success Factors (CPMSF) for sustainable social housing in Nigeria. *International Journal of Sustainable Built Environment, 3*(1), 62–71. doi:10.1016/j.ijsbe.2014.08.001

Kiani Mavi, R., & Standing, C. (2018). Identifying critical success factors for green business parks: Case study of Singapore. *Journal of Cleaner Production, 194*, 751–765. doi:10.1016/j.jclepro.2018.05.120

Korkmaz, S., Horman, M., Molenaar, K., & Gransberg, D. (2010). *Influence of project delivery methods on achieving sustainable high performance buildings report on case studies.* Research Sponsored by the Charles Pankow Foundation, DBIA. Retrieved from https://dbia.org/wp-content/uploads/2018/05/Research-Influence-on-Sustainable-High-Performance-Bldgs-Case-Studies.pdf

Korkmaz, S., Riley, D., & Horman, M. (2010). Piloting evaluation metrics for sustainable high-performance building project delivery. *Journal of Construction Engineering and Management, 136*(8), 877–885. doi:10.1061/(asce)co.1943-7862.0000195

Korkmaz, S., Riley, D., & Horman, M. (2011). Assessing project delivery for sustainable, high-performance buildings through mixed methods. *Architectural Engineering and Design Management, 7*(4), 266–274. doi:10.1080/17452007.2011.618675

Korkmaz, S., Swarup, L., & Riley, D. (2011). Delivering sustainable, high-performance buildings: Influence of project delivery methods on integration and project outcomes. *Journal of Management in Engineering, 29*(1), 71–78. doi:10.1061/(ASCE)ME.1943-5479.0000114

Lam, P. T., Chan, E. H., Poon, C. S., Chau, C. K., & Chun, K. P. (2010). Factors affecting the implementation of green specifications in construction. *Journal of Environmental Management, 91*(3), 654–661. doi:10.1016/j.jenvman.2009.09.029

Lei, Z. (2005). Is private finance initiative a good mechanism to deliver sustainable construction? Paper presented at the 2005 World Sustainable Building Conference, Tokyo, Japan.

Li, Y. Y., Chen, P.-H., Chew, D. A. S., & Teo, C. C. (2014). Exploration of critical resources and capabilities of design firms for delivering green building projects: Empirical studies in Singapore. *Habitat International, 41*, 229–235. doi:10.1016/j. habitatint.2013.08.008

Li, Y. Y., Chen, P.-H., Chew, D. A. S., Teo, C. C., & Ding, R. G. (2011). Critical project management factors of AEC firms for delivering green building projects in Singapore. *Journal of Construction Engineering and Management, 137*(12), 1153–1163. doi:10.1061/(ASCE)CO.1943-7862.0000370

Molenaar, K., Sobin, N., Gransberg, D., McCuen, T., Korkmaz, S., & Horman, M. (2009). *Sustainable, high performance projects and project delivery methods: A state-of-practice report.* Retrieved from http://citeseerx.ist.psu.edu/viewdoc/download?doi=10.1.1.537.8794&rep=rep1&type=pdf

Murtagh, N., Roberts, A., & Hind, R. (2016). The relationship between motivations of architectural designers and environmentally sustainable construction design. *Construction Management and Economics, 34*(1), 61–75. doi:10.1080/01446193. 2016.1178392

Olanipekun, A. O., Xia, B., Hon, C., & Darko, A. (2018). Effect of motivation and owner commitment on the delivery performance of green building projects. *Journal of Management in Engineering, 34*(1). doi:10.1061/(ASCE)ME.1943-5479.0000559

Oyebanji, A. O., Liyanage, C., & Akintoye, A. (2017). Critical success factors (CSFs) for achieving sustainable social housing (SSH). *International Journal of Sustainable Built Environment, 6*(1), 216–227. doi:10.1016/j.ijsbe.2017.03.006

Pheng Low, S., Gao, S., & Lin Tay, W. (2014). Comparative study of project management and critical success factors of greening new and existing buildings in Singapore. *Structural Survey, 32*(5), 413–433. doi:10.1108/SS-12-2013-0040

Rasekh, H., & McCarthy, T. J. (2016). Delivering sustainable building projects– challenges, reality and success. *Journal of Green Building, 11*(3), 143–161. doi:10.3992/jgb.11.3.143.1

Riley, D., Sanvido, V., Horman, M., McLaughlin, M., & Kerr, D. (2005). Lean and green: The role of design-build mechanical competencies in the design and construction of green buildings. Paper presented at the Construction Research Congress, San Diego, CA.

Robichaud, L. B., & Anantatmula, V. S. (2010). Greening project management practices for sustainable construction. *Journal of Management in Engineering, 27*(1), 48–57. doi:10.1061/(ASCE)ME.1943-5479.0000030

Shen, W., Tang, W., Siripanan, A., Lei, Z., Duffield, C. F., Wilson, D., ... Wei, Y. (2017). Critical success factors in Thailand's green building industry. *Journal of Asian Architecture and Building Engineering, 16*(2), 317–324. doi:10.3130/jaabe.16.317

Swarup, L., Korkmaz, S., & Riley, D. (2011). Project delivery metrics for sustainable, high-performance buildings. *Journal of Construction Engineering and Management, 137*(12), 1043–1051.

Tang, Z. W., Ng, S. T., & Skitmore, M. (2019). Influence of procurement systems to the success of sustainable buildings. *Journal of Cleaner Production, 218*, 1007–1030. doi:10.1016/j.jclepro.2019.01.213

Venkataraman, V., & Cheng, J. C. P. (2018). Critical success and failure factors for managing green building projects. *Journal of Architectural Engineering, 24*(4). doi:10.1061/(ASCE)AE.1943-5568.0000327

Wai, S., Yusof, A. M., Ismail, S., & Tey, K. (2012). Critical success factors for sustainable building in Malaysia. *International Proceedings of Economics Development and Research.* Retrieved from http://www.ipedr.com/

Wang, N., Wei, K., & Sun, H. (2013). Whole life project management approach to sustainability. *Journal of Management in Engineering, 30*(2), 246–255.

Xia, B., Skitmore, M., Wu, P., & Chen, Q. (2014). How public owners communicate the sustainability requirements of green design-build projects. *Journal of Construction Engineering and Management, 140*(8). doi:10.1061/(ASCE)CO.1943-7862.0000879

Zhang, L., Li, Q., & Zhou, J. (2017). Critical factors of low-carbon building development in China's urban area. *Journal of Cleaner Production, 142,* 3075–3082.

Zhou, L., & Smith, A. J. (2013). *Sustainability best practice in PPP: Case study of a hospital project in the UK.* Paper presented at the International Conference on PPP Body of Knowledge, Preston, UK.

Chapter 2

Green Building project success: Social conditions

2.1 Introduction

> **Who should read this chapter**
>
> Read this chapter if you are interested in understanding the social conditions enabling successful outcomes in Green Buildings (GBs).
>
> If you are interested in understanding the technical conditions enabling successful outcomes in GBs, see Chapter 3.

In GB projects, the eminent barriers are not the lack of technologies and assessment methods, but are instead the organisational and procedural difficulties involved in the adoption of new methods (Häkkinen & Belloni, 2011). For the successful implementation of GB projects, the non-technical aspects are equally important and often management factors determine the success or failure of a project (Li, Chen, Chew, Teo, & Ding, 2011). Accordingly, the objective of this chapter is to 'investigate the social success conditions of GB projects and their interrelationships'. In this chapter, the identified success conditions and their interrelationships are discussed within their nine broader themes. Overall, these themes respond to the critical role of all the key GB stakeholders in the successful development of GB projects. While addressing the themes, detailed accounts of the success conditions occurring within those themes are also provided. The interrelationships within the success conditions occurring in individual themes are also discussed. Upon reading this chapter, you would be able to understand the following:

- How some of the client's characteristics contribute to GB project performance
- How the cooperation and interest of key project stakeholders contribute towards GB project performance

DOI: 10.1201/9781003322740-4

- How the education of the project team, client, and end-users regarding the sustainable development and operation of a project affects GB project performance
- How the project team mindset, priorities, characteristics, authorities, responsibilities, contractual relationships, and commitment to the project affect GB performance
- The team procurement methodology contributing towards project performance

2.2 Cooperation and interest of stakeholders

'Cooperation and interest of stakeholders' are associated with GB project success. This *theme* has 12 conditions overall, among which the key ones include the following: 'Client's involvement in project development' ($n = 5$), 'Client's motivation to achieve sustainable outcomes' ($n = 32$), 'Cooperative role of building control authorities' ($n = 2$), 'End-users' operation of building in sustainable ways' ($n = 12$), and 'Stakeholders' approval of project' ($n = 4$). 'Client's involvement in project development', 'Client's motivation to achieve sustainable outcomes', and 'End-users' operation of building in sustainable ways' represent three *clusters* having one, three, and three sub-conditions, respectively. Conditions within this *theme* are related to the project client, end-users, building control authorities, and the neighbouring community (Table 5.1 and Figure 5.2).

According to the conditions represented by this theme, the likelihood of project success increases in case the client is directly involved in project sustainability and is engaged throughout the design process; endorses the sustainability brief; facilitates the coordination among project team members; has stakes in operational performance of the project; and mandates the tenants to follow sustainability guidelines. Alongside the interest and cooperation of project client, the GB project success is also determined by the cooperation of other stakeholders, for instance, cooperation and motivation of investors and building control authorities towards sustainable development, motivation and awareness of end-users to sustainably operate the building according to the design intent, and approval from the neighbourhood community to develop the project.

The importance of these conditions (as shown in Figure 2.1) for project success is that they result in client's awareness regarding the sustainable development and operation of building; ensure project funding to fulfil the sustainability goals; ensure the reduced impact of value engineering on sustainable features; and ensure that the project team takes the project sustainability goals seriously and is motivated to fulfil them. Most importantly, these conditions help achieve the sustainable operation of the building and enable the development of the project according to aspirations.

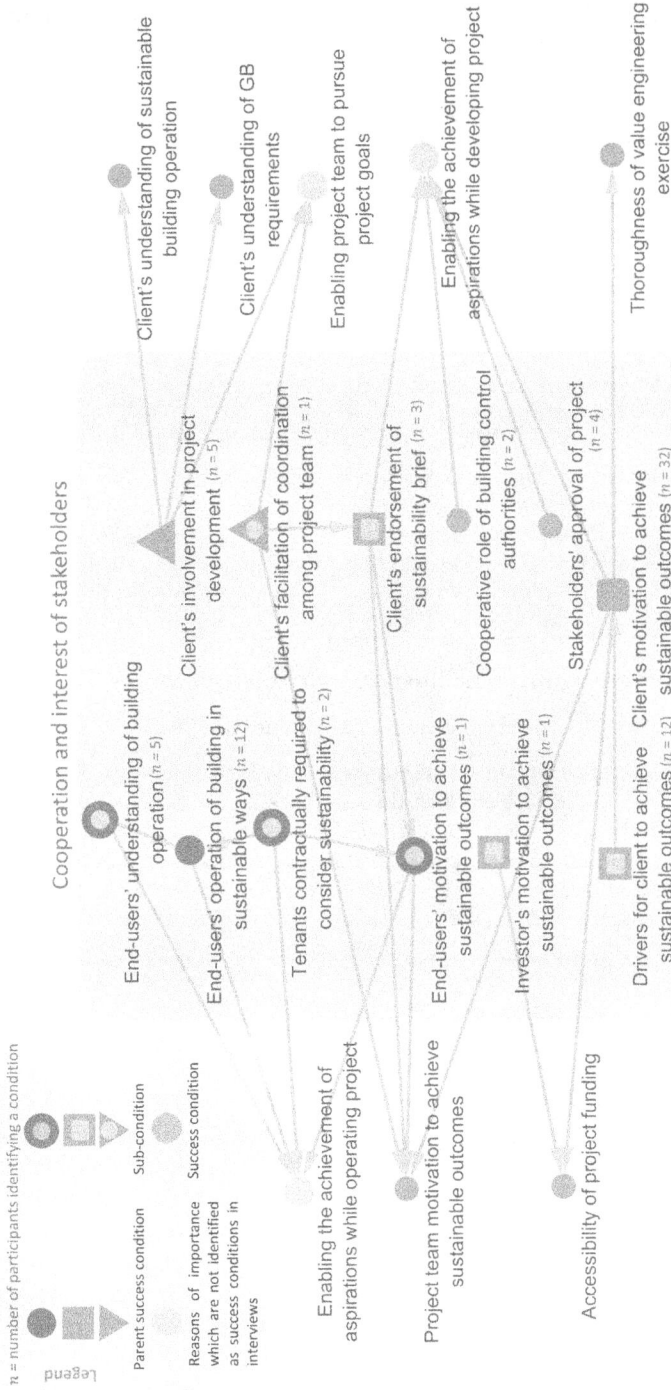

Figure 2.1 Conditions of 'Cooperation and interest of stakeholders'.

Conditions within this theme, as shown in Figure 2.1, can reduce the non-value-adding activities in a GB project by resulting in the client's understanding of sustainable building operation and GB requirements. These conditions can also help create value for the client by enabling the team to pursue project goals; enabling the achievement of aspirations while developing and operating a project; resulting in the motivation of client, end-users, and project team to achieve sustainable outcomes; facilitating the accessibility of project funding; resulting in the client's endorsement of sustainability brief; and avoiding the elimination of important building features during the value engineering exercise.

Client's involvement in project development

According to an Australia-based sustainability consultant (AU-M-19),

> The client or the client stakeholders need to be involved in the design process to know how things are intended to be operated. They will then know what they are going to get and how the building is going to work. They will know what they need to do to get the building working as designed.

Client's motivation to achieve sustainable outcomes

According to a UK-based engineering consultant (UK-M-4),

> Most of the clients are driven towards GBs for getting the planning approvals and the problem with this approach is that as soon as a problem arise in the project development, the client resorts to easier solutions which may not benefit the project sustainability performance, for instance many clients find it easier to pay the Carbon Tax. Many clients are not interested in making the building work, especially once the building is delivered.

2.3 Team characteristics

The team characteristics are important aspects associated with GB project success. This *theme* has 12 conditions (Figure 2.2) among which the key ones are 'Proficiency of project team' ($n = 32$), 'Project team's understanding of project goals and aspirations' ($n = 3$), 'Proficiency of FM team' ($n = 3$), 'Like-mindedness of project team members' ($n = 2$), and 'Size of design team' ($n = 2$). 'Proficiency of project team' is a condition representing a *cluster* with seven sub-conditions which are about leadership qualities in project team members ($n = 2$), and the proficiency of the contractor ($n = 12$), design consultant ($n = 12$), MEP consultant ($n = 6$), project management (PM) team ($n = 4$), sub-contractor ($n = 5$), and sustainability consultant ($n = 5$). Conditions within this *theme* are related to facility management (FM)

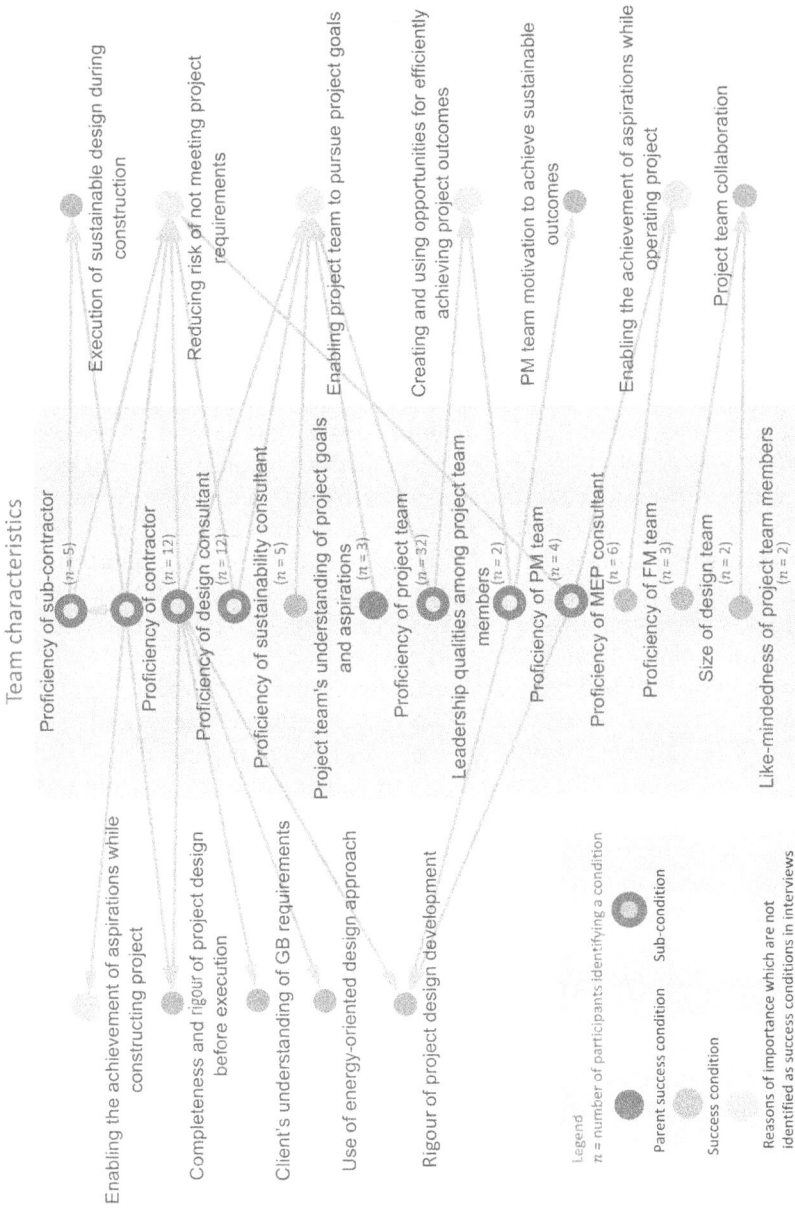

Figure 2.2 Conditions of 'Team characteristics'.

team, contractor team, design consultant, MEP consultant, PM team, sub-contractor, and sustainability consultant, as well as the overall project team (Table 5.1 and Figure 5.2).

According to the conditions represented by this theme, the likelihood of GB project success is associated with the differences in knowledge, thinking approach, and skill sets of project team members; the size of the design team; and the number of design consultant organisations involved in project development. The likelihood of GB project success increases in case: the FM team members have a good understanding of building features and building operation; the project team has a detailed understanding of project vision, goals, aspirations, and quality targets; and the project team is proficient to develop GB projects. Proficiency of project team implies that the project team is experienced in GB development; has the appropriate technical expertise required for GB projects; understands the process of GB development and certification; and has the leadership qualities to pursue project goals, push status-quo, and innovate. In specific terms, the proficiency of project team means that the sustainability consultant has the experience of GB development and the ability to engage with the project team; the design consultant has the understanding and expertise of GB development and certification; the design consultant is aware of local regulations and design requirements regarding efficient MEP systems; the MEP consultants have the experience of working on GBs, the experience of problem rectification in building systems, and the ability to make better design speculations; the project management team has the experience of delivering GBs, the ability to discern good and bad design, and can appropriately delegate the project tasks among the project team; the contractor has the awareness and experience of GB development, can bridge the knowledge gaps, understand design intent, review sustainability compliance by sub-contractors, and administer documentation related to GB certification; and the sub-contractors have the awareness and expertise of GB development.

The importance of these conditions for project success is that they ensure team collaboration; enable the team to work towards the achievement of project aspirations, work on innovative solutions, and focus on goals even in challenging circumstances; ensure that the project team can contribute towards the development and certification of a GB project; and help guarantee that the potential problems in building systems can be rectified once the project is operational. The importance of these conditions for project success is also because they ensure that the sustainability consultant can lead the project team to deliver sustainable outcomes; the MEP consultants can design the building systems which meet the sustainability aspirations; the design consultants can design the project rigorously, meeting sustainability aspirations in design, avoid the regulatory barriers, create conditions to achieve higher efficiencies in building systems, and compensate the client's lack of awareness regarding GB development; the contractor can monitor the deliverables from sub-contractors and suppliers for compliance, can compensate the lack of sub-contractors' expertise, can complete the documentation required for GB certification, and deliver the project aspirations even in case design deliverables are incomplete and flawed; the sub-contractors can deliver the project tasks assigned to them while meeting sustainability aspirations; the PM team can focus its attention towards fulfilling sustainability outcomes, can monitor and control the design development to meet the project aspirations, and ensure an appropriate delegation of responsibilities among suitable professionals; and the FM team can effectively operate the GB project.

Conditions within this theme, as shown in Figure 2.2, can reduce the non-value-adding activities in a GB project by resulting in project team collaboration, ensuring a complete and rigorously developed project design before the start of construction, and creating opportunities for efficiently achieving project outcomes. These conditions can also help create value for the client by enabling the achievement of aspirations while developing and operating a project; enabling the project team to pursue project goals; ensuring the rigorous design development; reducing the risk of not meeting project requirements; ensuring the client's understanding of GB requirements; and a rigorous execution of sustainable design during construction.

Proficiency of project team

While talking about a school building, a UK-based sustainability consultant (UK-F-4) mentioned,

> The project was aspired to achieve a BREEAM rating of Very Good. However, the project couldn't even achieve BREEAM rating of Good. In terms of energy performance, the project didn't meet its targets because of inefficient systems, inefficient fabric, etc.

> The project failed as it didn't have a good record keeping by the contractor responsible for project delivery. The team was not familiar with the BREEAM process. The evidence of the sustainable practices such as commissioning, and the material certificates couldn't be delivered to BREEAM assessors and therefore some credits were not awarded.

Proficiency of sub-contractor

According to a Singapore-based sustainability manager (SN-M-2),

> The main contractor often involves itself only with the building structure. Often the teams who work on the building systems and have the ability to impact the sustainability performance are the sub-contractors. If any sustainability-related trouble in a building is created, it is often due to the sub-contractors. These speciality contractors through years have developed their ways of doing things which help them save time and labour, but these ways can be in contradiction with sustainability.

Proficiency of project team

According to a Singapore-based engineering consultant (SN-M-7),

> Sometimes projects fail to meet their aspirations because the project teams are not much experienced in the installation of plants. The installation of every building plant is different and is not identical [to previous projects] because of the different heating and cooling loads. This is because the building occupants are different for every building. This strategizing and mitigation for each situation is not easy to come. Many professionals are not much experienced and don't have a lot of problem rectification experience. The professionals who have significant experience, can have a good understanding of the scenarios in which the HVAC system has to operate in future, so they design the system accordingly. Now a days, even in highly efficient buildings people complain that they feel hot or cold in the building. This is because the air conditioning may not be working properly. There are many problems of this nature which can be mitigated. It is important to note that this underperformance of the HVAC system may not be purely because of the design issue, it can also be because of the occupants doing the fit-out work in the building, and the fit-out contractor may not know about the do's and don'ts of the job. A lot of such issues can be mitigated if the landlord steps in to review the fit-out works, and if the landlord can ensure that the fit-out work will not affect the HVAC performance in case of alterations in space planning.

2.4 Team collaboration

The team-collaboration-related aspects are associated with GB project success. This *theme* has nine conditions (Figure 2.3) among which the key ones are 'Project team collaboration' ($n = 29$) and 'Conflicts among project team' ($n = 1$). 'Project team collaboration' is a condition representing a *cluster* with seven sub-conditions which are about team willingness to work together ($n = 4$), and liaison between the design team and client ($n = 5$), FM team and project development team ($n = 2$), sustainability consultant and client ($n = 1$), sustainability consultant and contractor team ($n = 3$), sustainability consultant and design consultant ($n = 4$), and sustainability consultant and project team ($n = 5$). Conditions within this *theme* are related to the client, FM team, contractor team, design consultant, and sustainability consultant, as well as the overall project team (Table 5.1 and Figure 5.2).

According to the conditions represented by this theme, the likelihood of GB project success increases in case the misunderstandings and conflicts among team members are avoided as the project develops and the project team collaborates for achieving aspirations. Team collaboration means that the project team is willing and excited to collaborate for the project; the team has effective working dynamics and starts collaborating from the project

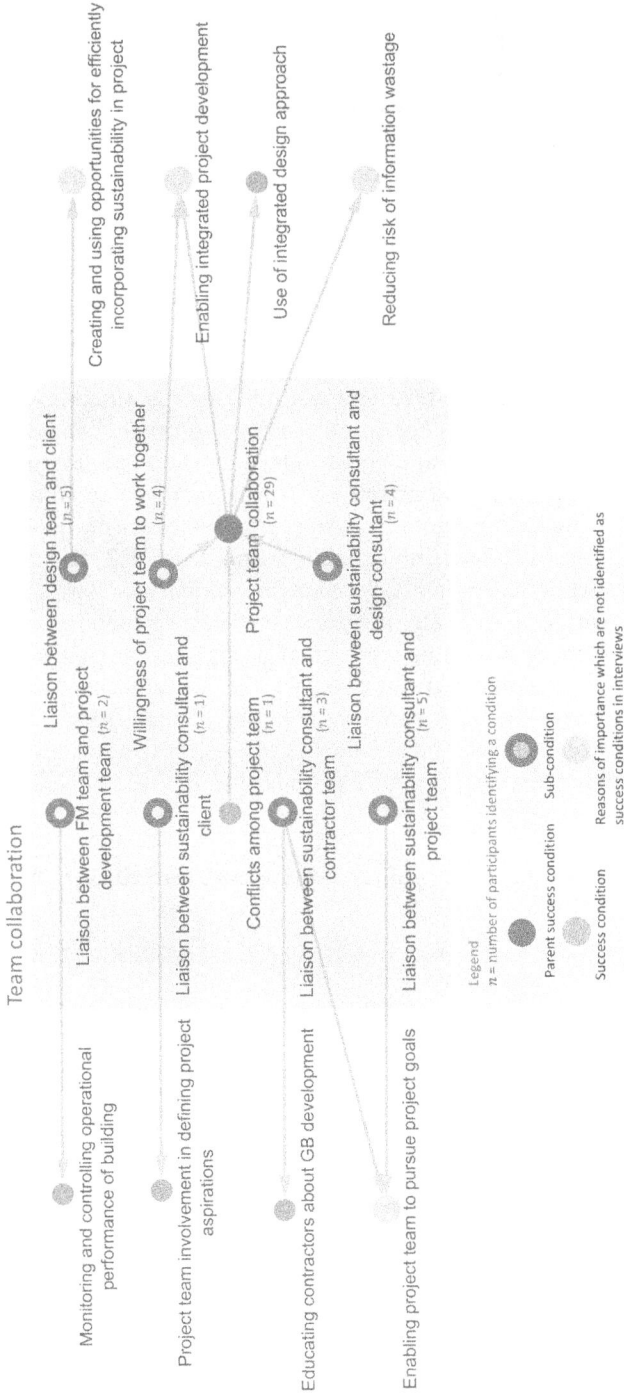

Figure 2.3 Conditions of 'Team collaboration'.

onset. Project team collaboration also implies that there is an active communication link between the design team and the client; sustainability consultant and the client; sustainability consultant and contractor team; architect, MEP consultant, and sustainability consultant; sustainability consultant and the overall project team; and the FM team and project development team.

The importance of these conditions for project success is that they enable the project team to work together for achieving project goals; help ensure that the team supports the project sustainability agenda; and help achieve the sustainable outcomes in the project. In specific terms, these conditions are important as they reduce the risk of design sub-optimisation and ensure an integrated design approach; help develop the client's trust in the design consultant, enabling the consultant to incorporate sustainable building features; enable a long-term assessment of building sustainability spanning from project development to operation; enable a mutual consensus among the sustainability consultant and client regarding the sustainability aspects for project development; and enable the sustainability consultant to lead the project team including the contractor team towards sustainability goals.

Conditions within this theme, as shown in Figure 2.3, can reduce the non-value-adding activities in a GB project by educating contractors about GB development; enabling an integrated project development; reducing the risk of information wastage; and creating opportunities for efficiently incorporating sustainability in the project. These conditions can also help create value for the client by enabling the project team to pursue project goals; ensuring team involvement in defining project aspirations; and controlling the operational performance of the building.

Liaison between sustainability consultant and design consultant

While talking about an office fit-out project, a UK-based sustainability consultant (UK-F-1) mentioned,

> The sustainability team did not get the opportunity to liaise directly with the design team. Although there were some meetings of them together but mostly the information between the two parties was shared through the project manager. The creation of this barrier and the separation of the sustainability team from the rest of the project team resulted in project failure. The project management team did not focus on the sustainability requirements of the building.

2.5 Team commitment to the project

The team commitment to the project is an important aspect associated with GB project success. This *theme* has ten conditions (Figure 2.4) among

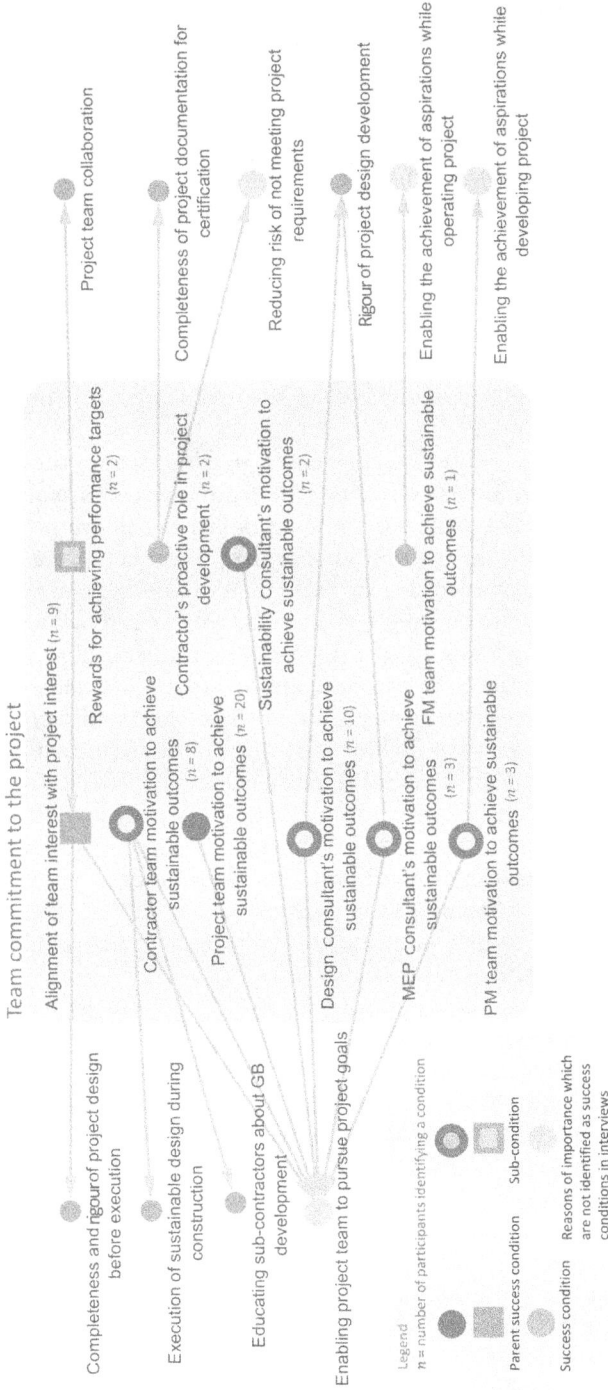

Figure 2.4 Conditions of 'Team commitment to the project'.

which the key ones are 'Project team motivation to achieve sustainable outcomes' ($n = 20$), 'Alignment of team interest with project interest' ($n = 9$), 'Contractor's proactive role in project development' ($n = 2$), and 'FM team motivation to achieve sustainable outcomes' ($n = 1$). 'Alignment of team interest with project interest' is a condition representing a *cluster* with one sub-condition, which is 'Rewards for achieving performance targets' ($n = 2$). 'Project team motivation to achieve sustainable outcomes' is also a condition representing a *cluster* with five sub-conditions which are exclusively related to the motivation of design consultant ($n = 10$), contractor team ($n = 8$), MEP consultant ($n = 3$), PM team ($n = 3$), and sustainability consultant ($n = 2$). Conditions within this *theme* are related to design consultant, contractor team, MEP consultant, FM team, PM team, and sustainability consultant, as well as the overall project team (Table 5.1 and Figure 5.2).

According to the conditions represented by this theme, the likelihood of GB project success increases in case: the team interests are aligned with project interest, and the project team is on-board to achieve project goals; there are incentives for the team to achieve the project goals; the contractor plays a proactive role by ensuring the buildability of design and avoiding the potential risks in the project; the FM team is motivated to achieve sustainable outcomes and operates a GB with special considerations; and there is interest and motivation of the project team towards sustainable outcomes. The motivation of project team implies that the sustainability consultant is committed to achieving sustainable outcomes; the sub-contractors, suppliers, head contractor team, and particularly the environmental manager in contractor team are cooperative and motivated towards sustainable development; design consultants and MEP consultants are motivated to deliver sustainable outcomes and are willing to develop the designs by accepting the recommendations from sustainability consultant; and the PM team takes interest in sustainable outcomes of the project and considers them as important aspects of project development.

The importance of these conditions for project success is that they help the project team to work collaboratively towards the project goals and enable the project team to maintain a focus on sustainability and strive for sustainable outcomes. More specifically, these conditions are important for project success as they enable the sustainability consultant to lead the project team towards sustainability objectives; enable the design consultants and MEP consultants to rigorously develop the project design which fulfils the sustainability aspirations; enable the contractor to deliver the project outcomes and ensure the completion of project documentation for green certification; and help the FM professionals to achieve the sustainable operational goals of the project.

Conditions within this theme, as shown in Figure 2.4, can reduce the non-value-adding activities in a GB project by resulting in project team collaboration; alignment of team interest with project interest; completeness and rigour of project design before execution; and awareness of contractor team about GB development. These conditions can also help create value for the client by resulting in the completeness of project documentation for certification; rigorous execution of sustainable design during construction; ease of project team to pursue project goals; reduced risk of not meeting project requirements; and the achievement of aspirations while developing and operating the project.

PM team and MEP consultant's motivation to achieve sustainable outcomes

While talking about an office building, a UK-based sustainability consultant (UK-M-5) mentioned,

> The client's representatives on this project especially the project manager had a non-serious attitude towards sustainability of the project. For instance, in the early project meeting relating the sustainability, the project manager was of the view that these issues didn't concern him and were perhaps a waste of his time. From the very beginning it was clear that a lot of effort was required in pushing the sustainability agenda forward. The project had a good architecture team. The building services team however, was not progressive and couldn't see the point of sustainability.

> The civil engineer and the services engineer questioned everything proposed by the sustainability team. They didn't want to help and acted as project sustainability was not part of their job. It happened even in the small things as they didn't even bother about the documentation for GB certification. Also, there was a lot of focus on reducing the project cost. To cut costs a lot of important design elements were Value Engineered out of the project.

MEP consultant's motivation to achieve sustainable outcomes

While talking about an Australia-based project, a sustainability consultant (AU-M-3) mentioned,

> In this project, the engineering consultancy found that there was no value in supporting the sustainability outcomes as the client was going to ignore those alternative proposals anyway. So instead of reinforcing the sustainability-related decision-making, the engineering consultants stood back, and it ultimately acted as another factor for the client to step further back from the sustainability related options. This was because the client found that instead of the whole project team only part of the team was supporting sustainability.

2.6 Client's characteristics

Associations of the characteristics of the client organisation with GB project success are identified. All the eight conditions identified in this regard are focused on the project client only (Table 5.1 and Figure 5.2). These conditions include 'Proficiency of project client' ($n = 20$), 'Structure and nature of client organisation' ($n = 5$), 'Client's leadership in project' ($n = 3$), and 'Consensus within client organisation' ($n = 2$). 'Proficiency of project client' also represents four sub-conditions, which are 'Client understanding the need of sustainable outcomes' ($n = 5$), 'Client's understanding of GB requirements' ($n = 12$), 'Client's understanding of sustainable building operation' ($n = 2$), and 'Client's rational decision-making' ($n = 1$).

According to the conditions represented by this theme, the likelihood of GB project success increases in case: the client uses leadership skills during project development, there is agreement among client team regarding project development, there are client's stakes in the operation of the project, and the client has proficiency in GB development. The proficiency of the project client implies that the client is experienced in developing GB projects or construction projects in general and can discern good and bad design; the client has a clarity that why sustainable outcomes are required, understands the requirements of developing GB projects and the sustainable operation of these projects; the client's decision-making is based on logical reasoning instead of prejudices; and the client has confidence in sustainable solutions and is willing to follow project guidelines.

Conditions related to client's characteristics, as shown in Figure 2.5, are important for project success as they enable rigour in the client's decision-making; motivate the client towards sustainable outcomes and towards the achievement of original project intent; ease the early availability of specialised project-related knowledge by early involvement of contractor team; enable early incorporation of sustainability and integrated project development; help define the project requirements; motivate the project team towards achieving sustainable outcomes; and help achieve the aspirations while developing and operating the project.

Some conditions in this *theme* can reduce the non-value-adding activities in a GB project by enabling an integrated project development; ensuring the early incorporation of sustainability in the project; and the provision of specialised knowledge of building development early in the project. All the conditions within this *theme* can help create value for the client and reduce the gap between achieved value and best possible value by motivating the client to achieve original project intent and sustainable outcomes; by the robust development of project goals, aspirations, and requirements; by enabling the achievement of aspirations while developing and operating a project; by motivating the project team

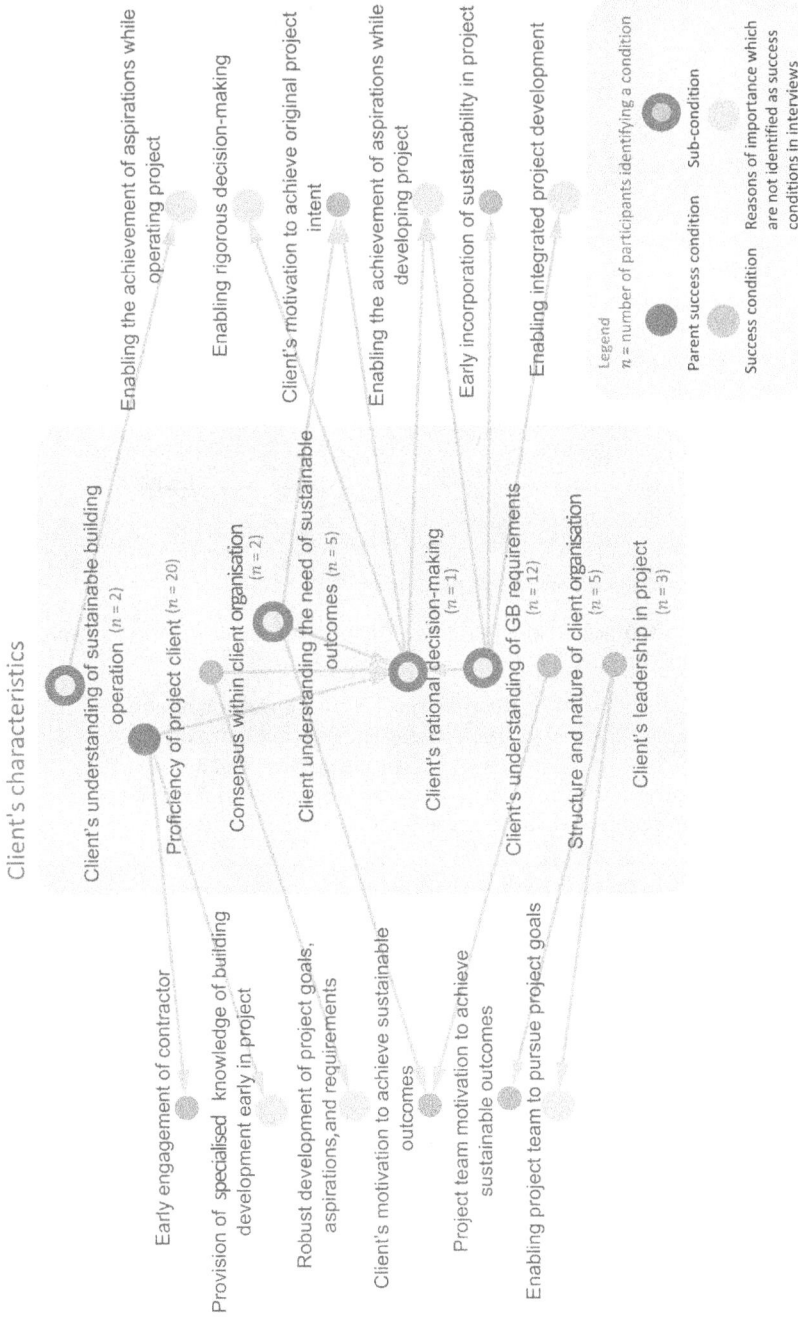

Figure 2.5 Conditions of 'Client's characteristics'.

to achieve sustainable outcomes; by enabling rigorous decision-making; and by enabling the project team to pursue project goals.

Client's understanding of GB requirements

While talking about a church building, a UK-based sustainability consultant (UK-M-1) mentioned,

> The client came back to us [sustainability consultant] and told us that the building was almost complete and asked us what was to be done to certify it with BREEAM. The client had not followed sustainability in design and construction, and had only followed the guidelines we provided at the project start, in a loose way. Certifications like BREEAM and sustainable project development in general depend on the process used for developing a Green Building and is not merely about parachuting yourself into sustainability as the project finishes and asking for BREEAM certification.

Consensus within client organisation

While talking about a housing project in Australia, a sustainability consultant (AU-M-1) mentioned,

> The project was a JV of the developer with a government entity which made the case of higher capital spending for photovoltaics, a challenge. In case of JVs, each party shares half of the project, and it is very difficult to agree on innovations. JV parties can agree to standard approaches but it is difficult for them to agree on innovations.

Client's rational decision-making

While talking about an office building project in Australia, a sustainability consultant (AU-M-3) mentioned,

> In this project, there were more benefits achieved than the failures incurred, however, many potential opportunities were left untapped. The project was being developed by an organization who cared a lot about what others will say, rather than the project outcomes. They made emotional decisions rather than rational ones. Detailed analyses would suggest one thing, but the client would opt for another thing. They couldn't look beyond their own prejudices. The project could not achieve some potential benefits and outcomes because the client made some decisions outside the project guidelines they had established earlier in the project.

While talking about a large-scale project in Australia comprised of multiple specialised facilities, a sustainability consultant (AU-M-3) mentioned,

> When making decisions regarding building materials the client would default back and say that they haven't done that before and therefore wouldn't do it on that project. We as sustainability consultant engaged with the architect to look for the materials that increase sustainability credentials of the building, provide better indoor environment, etc. These were the aspects of materials that the client was particularly interested in, as their previous projects lacked in this regard. However, when the client found that the new alternatives were significantly different from the previous materials used on their previous projects, they stood back into their shell and used the same materials they had used before. The client ended up facing the same problems as in the previous projects.

2.7 Team mindset and priorities

The team mindset and priorities are important aspects associated with GB project success. This *theme* has six conditions (Figure 2.6) which are 'Priority of sustainability in project development' ($n = 9$), 'Team working on project with value management mindset' ($n = 5$), 'Open-mindedness and flexibility of project team' ($n = 3$), 'Establishing and promoting synergies' ($n = 2$), 'Focus of sustainability consultant on project goals' ($n = 1$), and 'Team working on project with innovative mindset' ($n = 1$). Conditions within this *theme* are related to the client and sustainability consultant as well as the overall project team (Table 5.1 and Figure 5.2).

According to the conditions represented by this theme, the likelihood of GB project success increases in case; synergies among project team are established and promoted during the course of project development; the project team works with a value management mindset and explores suitable economic solutions to ensure cost-benefit optimisation; the project team has the flexibility to adapt to the GB project requirements; and the project team has an innovative mindset and does not consider the project in terms of business as usual. Moreover, the sustainability consultant has a focus on project goals and can prioritise important aspects in the project; the design team members have the open-mindedness to accept each other's opinions; and sustainability is prioritised in project agenda and the guidelines from the sustainability consultant are seriously followed.

These conditions are important for project success as they result in increased value for different project features; enable the project team to deliver aspired sustainability outcomes in project development; result in effective solutions and approaches to sustainable project development; reduce the risk of eliminating project features in the later stages of project development; enable better communication among design team; and drive the project team to achieve sustainability in project development.

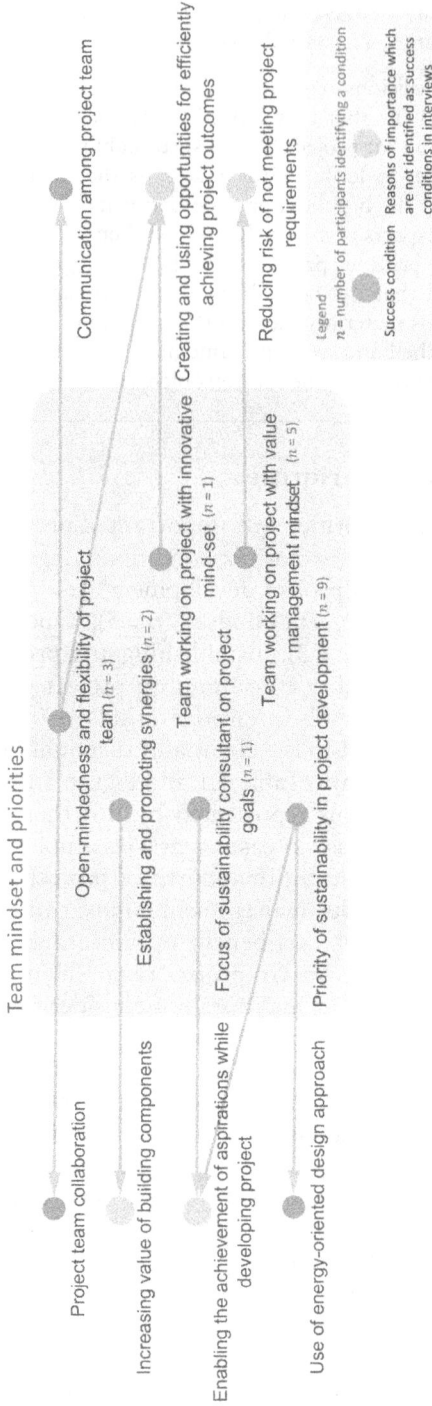

Figure 2.6 Conditions of 'Team mindset and priorities'.

Conditions within this theme, as shown in Figure 2.6, can reduce the non-value-adding activities in a GB project by resulting in communication and collaboration among the project team and creating opportunities for efficiently achieving project outcomes. These conditions can also help create value for the client by reducing the risk of not meeting project requirements; enabling the achievement of aspirations while developing the project; and increasing the value of building components. One condition, that is, 'Team working on project with value management mindset' is also related to the transformation of the project.

Priority of sustainability in project development

While talking about a project, an Australia-based sustainability consultant (AU-F-6) mentioned,

> In case of a failed project, the aspiration was to achieve five Star Green Star rating. However, as the project was being designed there was a lot of disagreement because the client was much heavily driven towards architectural design and did not agree to change some of the architectural design features including façade to achieve sustainability objectives. The client didn't look at the sustainability as integral to other services and to architecture. The client was very much drawn towards architectural features and when the architect would suggest some element, the client was driven to accept it even though they were warned by sustainability consultant that it will significantly increase the building operational cost. It appears that the project brief though containing high sustainability aspirations did not comply with the actual vision of the client and therefore the established sustainability objectives couldn't be fulfilled in that project.

While talking about a learning and teaching facility in Australia, a project manager (AU-M-8) mentioned,

> The reason for low performance in sustainability was that the architectural aspects were considered more ambitiously than the environmental aspects. Secondly, the project did not have enough time in the programme to achieve ambitious levels of sustainability. In this project, the lack of time instead of the lack of funds was the major reason that the ambitious sustainability goals could not be pursued.

According to an Australia-based sustainability consultant (AU-F-6),

> Generally, the building professionals are quite collaborative in the Melbourne market when they understand sustainability as an important project outcome. In case sustainability is a main driver for the project, the consultant team has no choice than to care.

According to a Hong Kong-based design consultant (HK-M-5),

> The pace of construction in Hong Kong is quite fast. Environmental design is not like structural or building services design, since it is only an optional approach in Hong Kong. In case the developer or contractor is posing pressure on design consultants, the time and cost requirements take a priority and the environmental design maybe compromised during the whole process.

2.8 Education

The education of the project team and key stakeholders (such as client, end-users, and FM team) regarding the importance of sustainable project outcomes and the process of GB development is an important aspect associated with GB project success. Among the five conditions occurring in this *theme* (Figure 2.7), the key ones are 'Educating client about sustainability in project' ($n = 6$), 'Educating end-users and FM team about building operation' ($n = 9$), and 'Educating project team about GB development' ($n = 8$). Conditions within this *theme* are related to the project client, end-users, FM team, project team, and contractor team (Table 5.1 and Figure 5.2).

According to the conditions represented by this theme, the likelihood of GB project success increases in case: the client is educated about the project sustainability as well as the tangible (such as ROI) and intangible benefits of sustainable development; the client, end-users, and FM team members are educated about the sustainable operation of building; and the project team is educated about GB development. Educating project team implies that the orientation of project team is conducted to bring all members on the same page of understanding related to GB development and GB certification systems, and the project team, and particularly the contractor team, is educated about GB development and made aware of their role in the delivery of sustainable outcomes.

These conditions, as shown in Figure 2.7, are important for project success as they ensure that the client sees value in sustainable development, is well informed to make rational decisions, and supports the project team for the sustainable development of the project; the FM team and end-users understand the requirements and process of sustainably operating the building; a smooth transition of the project takes place from inception to operation stage; the project team members and particularly the contractors and sub-contractors know about the process of GB development and their role in such development; and the entire team is on the same page of understanding related to GB development.

Conditions within this theme, as shown in Figure 2.7, can reduce the non-value-adding activities in a GB project by contributing towards the proficiency of the contractor, sub-contractor, and the overall project team;

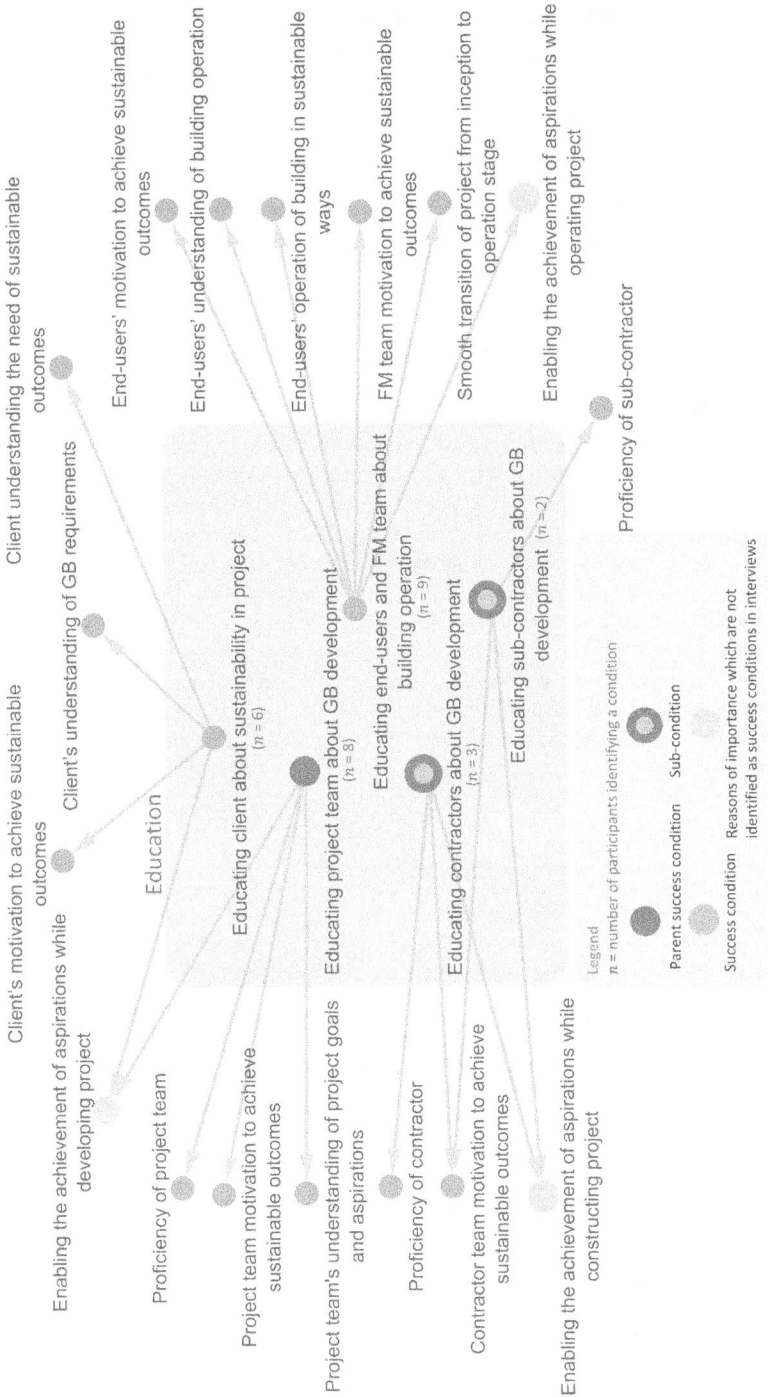

Figure 2.7 Conditions of 'Education'.

contributing towards the project team's understanding of project goals and aspirations and client's understanding of GB requirements; and a smooth transition of the project from inception to operation stage. These conditions can also help create value for the client by enabling the achievement of aspirations while developing and operating the project; motivating the client, end-users, FM team, contractor team, and the overall project team to achieve sustainable outcomes; contributing towards the client's understanding of the need of sustainable development; and resulting in end-users' understanding of building operation.

Educating contractors about GB development

According to a Hong Kong-based design consultant (HK-M-7),

> How to engage the contractor, as well as monitor and supervise the building construction are the key factors in avoiding failures in Green Building projects .... Once you get a design which meets the target on paper, the biggest challenge is to execute it. You must ensure that the contractor understands what you are targeting for. Some contractors in Hong Kong are already quite familiar with Green Building requirements and some are not. So, for some cost-conscious and tight-budget projects, the best contractors may not be engaged. This way in some cases we have to engage the contractors who may not understand Green Buildings well. In the process we have to educate them and ensure that they are well monitored. Further, during the construction, we have to ensure that their learning curve is kept up. These are the main reasons that some projects could not achieve their green targets.

Educating client about sustainability in project

According to an Australia-based sustainability manager (AU-M-15),

> For sustainable residential developments, the consultants often are in link with the developers and not the final owners of the property. Hence, reducing the opportunity for the consultant to educate the final owner. In case of the commercial clients, the education process is crucial because such projects also need to perform well operationally, for instance in terms of NABERS rating. Hence, the sustainability consultants need to train the clients in terms of proper usage of the building. The consultants are educating the clients informally throughout the project development, but the formal education happens closer to the project hand over stage. This is crucial to ensure that the project is operated sustainably.

According to a UAE-based sustainability consultant (AE-M-2),

> Around 50% of the clients bring the sustainability concept late in the project design stage. This is because of the lack of client's awareness. These clients think that LEED certification is an additional product they can get for the project at the end of delivery. Clients don't think that LEED certification is a whole process that needs to be started at the early design stages. They think of it as a marketing tool which they can buy while finishing the construction. Sometimes they think that they should have the design completed so that the sustainability consultants can provide assessments or give recommendations. Developers sometime also do the same mistakes.

Educating end-users and FM team about building operation

According to a UK-based engineering consultant (UK-M-4),

> There are many examples of projects in which failure occurred because of the unexpected circumstances and the lack of understanding of client's requirements. Many of the problems in such projects occurred once the buildings were occupied. This was because building engineers didn't give a thought that who will use the building, how will the building be used, and do the occupants understand how to use the building. So, the issue is that the transfer of building's Know-How doesn't happen. This issue is sometimes addressed through Soft Landings process in which depending on the complexity of a building, a short period is allocated during which the building designer briefs the facility manager how to run the building. But the problem is that after this period the circumstances can change a lot. For instance, the company occupying the building may leave and a new company may take over and the facility management-related knowledge of the building might be lost.

According to an Australia-based design consultant (AU-M-2),

> With new technology, it takes a longer time for the people to get familiar with the working of different systems. A Green Building may take up to one year, i.e., a full cycle of seasonal change to settle down. The problem is that this is an awfully long time for the occupants before they are comfortable. During this period there are a lot of complaints from the occupants, and they may desire to revert to the systems they know because they think that the systems in the building don't work. As the time passes, people get familiar with the building and understand it and the complaints begin to drop. A similar thing happened for the faculty of engineering building at UTS. They got complaints from users as the users didn't know how to run the building. The training in this case significantly helped the matters.

Failure of communication is probably one of the biggest problems. When a green building is handed over, it can have a number of innovative systems in it. In case those systems are not properly explained to maintenance or the facilities management team then it could mean that those systems may not work properly and get shut down eventually. A classic example is that half of the trigeneration systems in Sydney got shut down after 2-3 years of installation as people got fed-up with them and found that it is easier to get their electricity from the grid. This is probably because a system is not fully understood when it is designed, particularly when it is at the cutting edge of the technology. This means that the systems may not behave in an environment the way you want it to. This can be dealt by monitoring, understanding and fine tuning the building. This happens as you give people a new technology when they are used to conventional technology. It is similar to the example that you give a Formula One racing car to someone using a Holden. Unless there is a lot of training involved, the person new to the technology is likely to have trouble with it.

2.9 Team procurement methodology

The team procurement methodology is an important aspect associated with GB project success. This *theme* has seven conditions (Figure 2.8) among which the key ones are 'Preferences in project team selection' ($n = 12$) and 'Involvement of sustainability consultant in contractor's selection' ($n = 1$). 'Preferences in project team selection' is a condition representing a *cluster* with five sub-conditions which are 'Long-term engagement of sustainability consultant' ($n = 2$), 'Preferences in consultant's engagement' ($n = 5$), 'Preferences in contractor's engagement' ($n = 5$), 'Pre-qualification of contractors and sub-contractors' ($n = 1$), and 'Requirement for contractor to engage a sustainability advisor' ($n = 1$). Conditions within this *theme* are related to design consultant, contractor team, and sustainability consultant as well as the overall project team (Table 5.1 and Figure 5.2).

According to the conditions represented in this theme, the likelihood of GB project success increases in case the project team is carefully selected based on some criteria including the behavioural aspects; the sustainability consultant is engaged for the whole duration of project development and is involved in contractor's selection; a pre-qualification process of contractors and sub-contractors is conducted; the contractor team is required to have a sustainability advisor; the contractors selected are motivated to work on GBs, can adapt themselves to project requirements, and have the experience, knowledge, and skill set necessary for GB development; and the design consultants are carefully selected considering their ability to work on GB projects and their ability to adapt themselves to the project requirements. The importance of these conditions for project success is that they help engage a project team proficient in GB development, which can

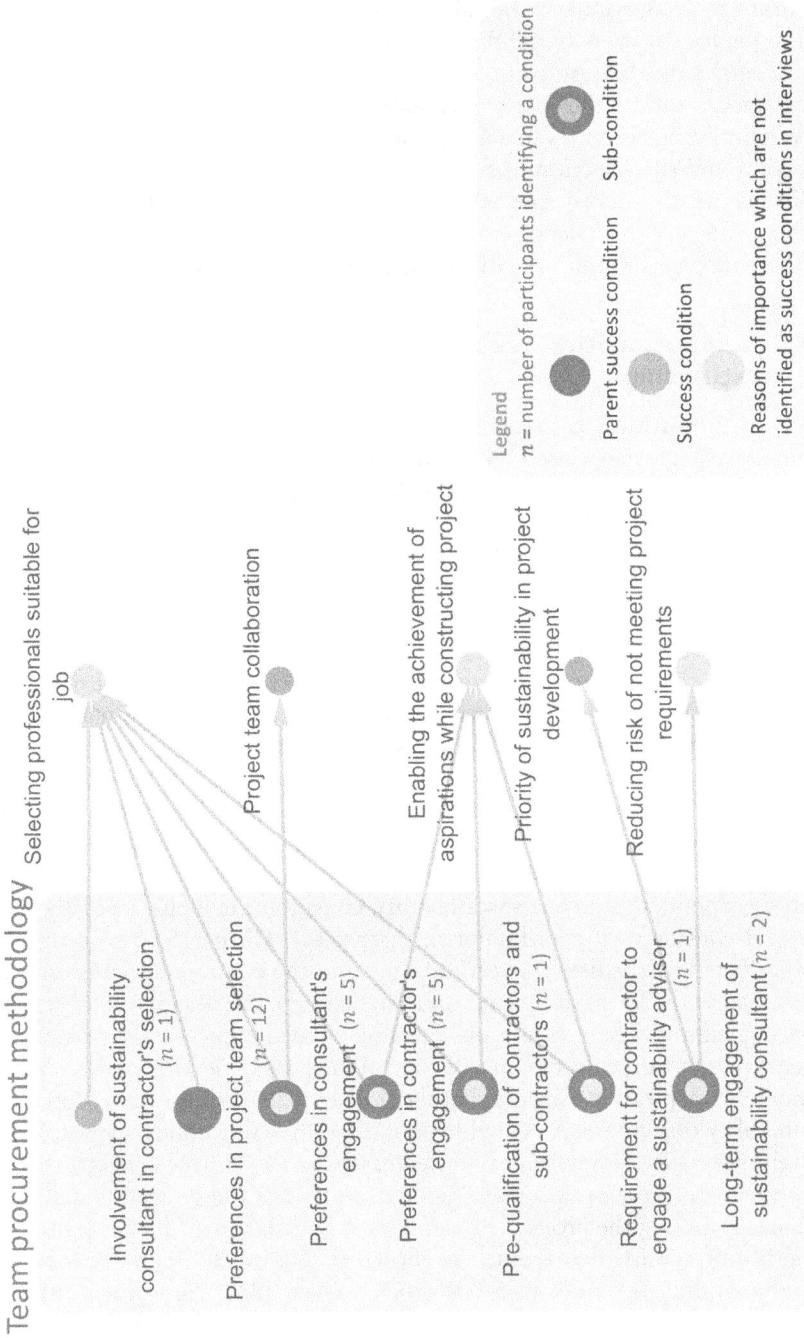

Figure 2.8 Conditions of 'Team procurement methodology'.

effectively collaborate for delivering the project; enable the sustainability consultant to incorporate sustainability aspirations in the project; ensure a continuous focus on sustainability as the project is developed; and reduce the risk of not meeting project aspirations.

Conditions within this theme, as shown in Figure 2.8, can reduce the non-value-adding activities in a GB project by resulting in project team collaboration and the selection of suitable professionals for the project. These conditions can also help create value for the client by reducing the risk of not meeting project requirements; enabling the achievement of aspirations while constructing project; and prioritising sustainability in project development.

2.10 Team authorities, responsibilities, and contractual relationships

The team authorities, responsibilities, and contractual relationships are important aspects associated with GB project success. This *theme* has five conditions (Figure 2.9) among which the key ones are 'Using appropriate project delivery method' ($n = 3$), 'Empowerment of sustainability consultant by client' ($n = 3$), 'Control of project design by Design and Sustainability Consultant' ($n = 2$), and 'Project team's involvement in decision-making' ($n = 2$). 'Using appropriate project delivery method' ($n = 3$) is a condition representing a *cluster* with one sub-condition which is 'Contractual interrelationships between client and project team' ($n = 2$). Conditions within this *theme* are related to the client, sustainability consultant, design team, and the overall project team (Table 5.1 and Figure 5.2).

According to the conditions represented by this theme, the likelihood of GB project success increases in case: the project team and particularly the sustainability and design consultants are involved in project-related decision-making; the design and sustainability consultants have the authority to control project design development; and the sustainability consultant is empowered by the client and considered a coordinator in the project. GB project performance is also affected by the delivery method used and the contractual links which a client has with the project team and particularly the design and sustainability consultant. These conditions are important for project success since they ensure that the project team can control the project development to meet sustainability aspirations; ensure that the project team seriously considers the sustainability objectives advocated by sustainability consultant; help capitalise on the specialised knowledge of different team members for benefitting the project; help keep the project team motivated towards the project; and affect associations among the project team and client which benefit the project.

Conditions within this theme, as shown in Figure 2.9, can reduce the non-value-adding activities in a GB project by enabling rigorous decision-making and aligning team interest with project interest. These conditions can also help create value for the client by resulting in rigorous design

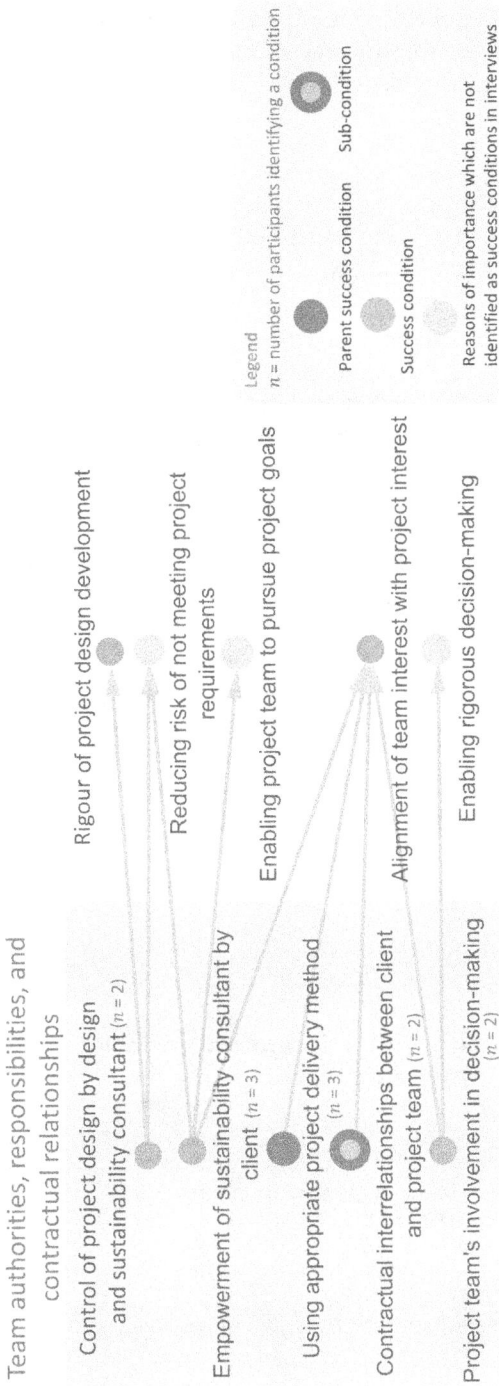

Figure 2.9 Conditions of 'Team authorities, responsibilities, and contractual relationships'.

development; enabling the project team to pursue project goals; and reducing the risk of not meeting project requirements. One condition, that is, 'Using appropriate project delivery method', is also related to the transformation of the project.

2.11 Summary

This chapter is the first among the two chapters to provide a detailed account of GB success conditions identified by 75 industry experts from Australia, the UK, the UAE, Singapore, Hong Kong, and Pakistan. This chapter addressed the stakeholder-oriented and non-technical success conditions of GB projects. Together, these attributes constitute nine broader themes comprised of 35 success conditions and 39 sub-conditions. This chapter discussed the client's characteristics; cooperation among project stakeholders and their interest in the project; education of project team, client, and end-users regarding sustainable development and use of GBs; team authorities, responsibilities, and contractual relationships contributing to project performance; team mindset, priorities, characteristics, collaboration, and commitment to the project; and team procurement methodology. The success conditions indicate that GB project performance is dependent on the interest of the project team and decision-makers in sustainable performance, their understanding of the process of sustainable project development, and their capability to steer the project towards aspired goals. These success conditions also indicate the approaches to improve cohesion and collaboration among project stakeholders, increase their awareness about GB development and operation, and stimulate in them a drive towards the vision of successful project development. The next chapter is about the technical conditions contributing to the success of GB projects.

References

Häkkinen, T., & Belloni, K. (2011). Barriers and drivers for sustainable building. *Building Research & Information*, *39*(3), 239–255. doi:10.1080/09613218.2011.561948

Li, Y. Y., Chen, P.-H., Chew, D. A. S., Teo, C. C., & Ding, R. G. (2011). Critical project management factors of AEC firms for delivering green building projects in Singapore. *Journal of Construction Engineering and Management*, *137*(12), 1153–1163. doi:10.1061/(ASCE)CO.1943-7862.0000370

Chapter 3

Green Building project success: Technical conditions

3.1 Introduction

> **Who should read this chapter**
>
> Read this chapter, if you are interested in understanding the technical conditions enabling successful outcomes in Green Building (GB) projects. If you are interested in understanding the social conditions enabling successful outcomes in GBs, see Chapter 2

The majority of construction projects is still being carried out according to the traditional methods, with short-term solutions preferred in place of long-term solutions and with the incorporation of technical approaches and material selection which can hardly be regarded as innovative green practices (Demaid & Quintas, 2006; Gluch, Gustafsson, & Thuvander, 2009; Hwang & Tan, 2012). To ensure an increase in the number of GB projects and to ease the achievement of sustainability aspirations, more attention is required towards the technical aspects of their development and delivery. Accordingly, this chapter aims to address the technical success conditions of GB projects. In this chapter, the technical success conditions identified in the semi-structured interviews with experts are discussed within 11 themes. While addressing the themes, detailed accounts of the success conditions occurring within those themes are also provided. The interrelationships within the success conditions occurring in individual themes are also discussed. Upon reading this chapter, you would be able to understand the following:

- How to manage changes during project development and fulfil design intent for better project performance
- Need for clarity in project development to ensure project performance
- Need of defining project goals and having complete and rigorously identified project deliverables

DOI: 10.1201/9781003322740-5

- Effects of constraints such as funding, logistics, and material availability on project performance
- Myriad of factors affecting project design and the approaches for rigorous design development
- Need and means of engraving sustainability in the development process to ensure project performance
- Need and means of performing project activities in a timely manner
- Flow of project information to ensure better project performance
- Inspection, monitoring, and control required to achieve high performance in projects
- Planning approaches that lead to high performance in projects

3.2 Design methodology

The approaches adopted in design development are associated with GB project success. Among the 18 conditions occurring in this *theme* (Figure 3.1), the key ones are 'Market survey to identify successful building systems' ($n = 1$), 'Use of performance-based specifications' ($n = 1$), and 'Rigour of project design development' ($n = 41$). Conditions within this *theme* are mostly related to the design team and in some cases also related to the project client and contractor team (Table 5.1 and Figure 5.2).

According to the conditions represented by this theme, the likelihood of GB project success increases in case the successful building systems available in the marketplace are identified, performance-based equipment specifications are used in procurement, and the project design and specifications are rigorously developed. Rigour of design development implies a consideration towards a life-cycle-based, energy-oriented, context-oriented, integrated, holistic, and innovative design approach. It means a design consideration towards intended functional use of the building, adaptability of building to multiple natures of usage, reduced complexity, building maintainability, and constructability. Rigour of design development also means the use of reliable technology and solutions; proactive design approach which is about avoiding the practice of making mistakes and then correcting them later; balanced design approach which considers mutually contradicting design objectives; and the approaches which reduce the gap between the speculated design conditions and actual conditions. The importance of these conditions for GB project success is that they reduce the risk of not meeting aspirations; create opportunities to efficiently incorporate sustainability and achieve outcomes while developing and operating the project; enable the project team to pursue project goals; enable integrated project development; increase the value of building components; and reduce the risk of scope changes during execution.

Conditions within this theme, as shown in Figure 3.1, can reduce the non-value-adding activities in a GB project by creating opportunities for

Enabling project team to pursue project goals

Scope changes during project execution

Reducing risk of not meeting operational aspirations of project

Legend
n = number of participants identifying a condition   Sub-condition

Parent success condition

Success condition   Reasons of importance which are not identified as success conditions in interviews

Creating and using opportunities for efficiently incorporating sustainability in project

Increasing value of building components

Enabling integrated project development

Creating and using opportunities for efficiently achieving project outcomes

Design methodology

Use of performance-based specifications (n = 1)

Suitability of project design for execution (n = 3)

Use of energy-oriented design approach (n = 12)

Use of holistic design approach (n = 9)

Maintainability considered in building design (n = 3)

Speculation in building design (n = 7)

Adding adaptability and multiple layers of use in building design (n = 1)

Design consideration towards future building usage (n = 3)

Use of reliable technology and solutions (n = 2)

Use of balanced design approach (n = 1)

Use of integrated design approach (n = 13)

Rigor of project design development (n = 41)

Use of context-oriented design approach (n = 3)

Life-cycle-based project development approach (n = 4)

Use of innovative design approach (n = 6)

Use of a proactive design approach (n = 5)

Level of complexity in project design (n = 4)

Market survey to identify successful building systems (n = 1)

Enabling the achievement of aspirations while constructing project

Enabling the achievement of aspirations while operating project

Reducing risk of not meeting project requirements

Enabling the achievement of aspirations while developing project

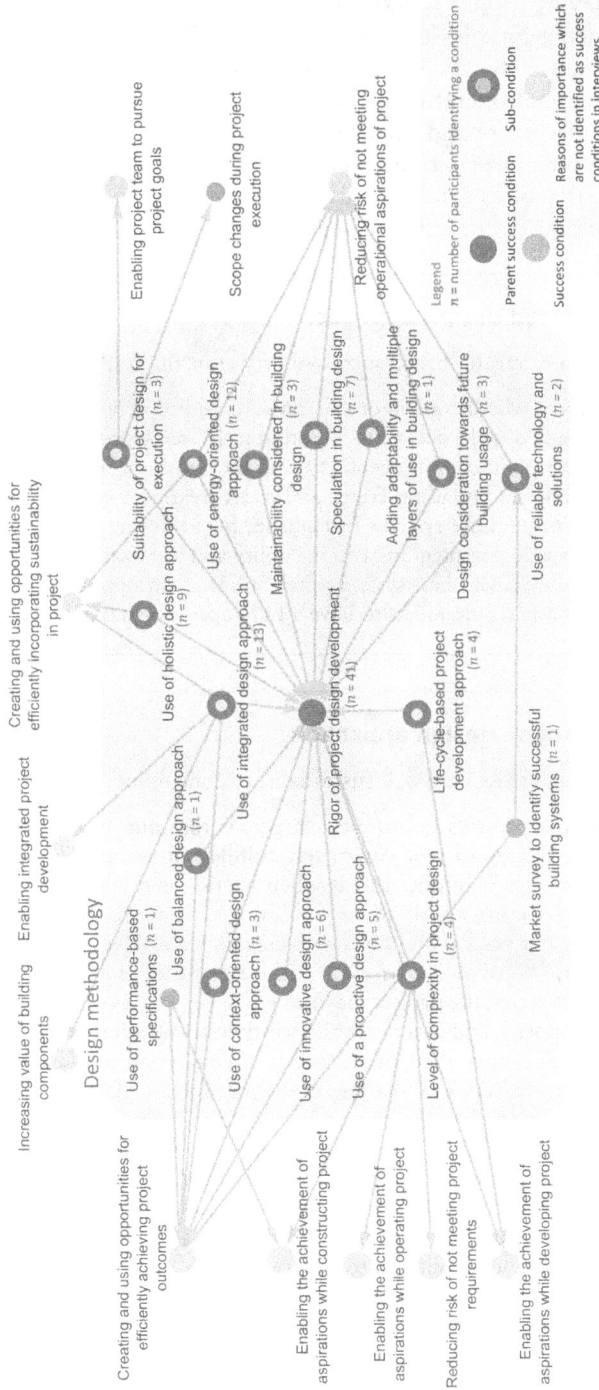

Figure 3.1 Conditions of 'Design methodology'.

efficiently achieving project outcomes, enabling integrated project development, and reducing complexity in project design. These conditions can also help create value for the client by enabling the achievement of aspirations while developing and operating the project; reducing the risk of not meeting project requirements and operational aspirations; increasing the value of building components; enabling the team to pursue project goals; creating and using opportunities for efficiently incorporating sustainability in the project; and helping avoid scope changes during project execution.

Use of integrated design approach

According to an Australia-based sustainability consultant (AU-M-19),

> Any initiatives whether green or not, have to be properly integrated in the built form and should not act as an add-on. In many green buildings, the clients have a point chasing mentality, and they go for a bolt-on approach. These buildings end up using Trigen and water treatment systems to achieve points. But often those systems brought in, for achieving credit points are flawed as they do not integrate with the utility and functionality of the building. As a consequence, these systems often fail. So, such a project fails to meet the client's basic requirements to have a fully operational building.

Use of a proactive design approach

According to a Hong Kong-based sustainability consultant (HK-M-2),

> A reason why projects fail to meet targets is that our aspirations in the engineering design are not completely fulfilled in reality. The most frequent occurrence regarding this is when we do energy modelling. There are so many factors which affect building performance. For instance, in Hong Kong most of the buildings with curtain walls have a high window-to-wall ratio. We would advise to reduce this ratio. However, when the construction starts, the project may not get developed as per the solutions initially proposed, and as a result the window-to-wall ratio may increase. The clients also have their preference in the selection of glazing colour. We cannot use a very low shading coefficient for the glass as the natural day light usage needs to be ensured in case of green buildings. If the glass does not let enough light through it, the indoor environment will be dim. We often propose several solutions to the client, for instance, changing glass and adding vertical louvers. However, still during the construction the client goes for variations in the design. In such situations, we start thinking about other options to secure enough points for meeting certification targets. For instance, we begin thinking about the use of PV panels or biodiesel generators to compensate the drawbacks from other low performing design aspects such as building façade.

Speculation in building design

According to a Singapore-based sustainability consultant (SN-M-11),

> Projects may suffer from some failures because of the misinterpretation of information. We base the design on data collection. The data relates to different building typologies, but it may not be perfect and when you design using this data, you will not know of the building performance until the building is operational and the energy consumption of the building is stabilised.

According to a Singapore-based engineering consultant (SN-M-6),

> Failures also happen when the consultants are too much speculative in their designs. For instance, in one of our projects, the client required us to design a very heavy air-conditioning system. The client speculated that the building would require a big system because of the potential use of high-power equipment in building's operational life. However, in reality the high-power equipment was never used in the building and the air-conditioning being too big for the building, resulted in mold, which damaged the carpet and the ceiling. Along with the property damage, such flaws in design considerations can also lead to sickness of building-users and loss of worker productivity. So, when there is a lack of clarity in the project brief, it becomes a major reason for failure.

Design consideration towards future building usage

According to a UK-based engineering consultant (UK-M-4),

> Many things which work in theory, don't work in practice. Sometimes to achieve as many as possible Green credentials for a building, some combinations of systems are used which may not work well in unison. A typical example is of a Combined Heat and Power system in which you [may] end up creating a lot of thermal energy stored in the buffer tanks. In case you don't have a mixed use, you may end up having more energy than required and, in that case, you may need radiators to discharge extra heat. This can occur because of various reasons including the miscalculations from the design team and the unexpected issues. The unexpected issues can come in many forms, for instance, a project may not be occupied as it was meant to, and there could be economic issues involved as the credit crunch of 2009. So, the major issue is that the planning for a project is not coherent with the reality, and it can result in building plants being oversized.
>
> There is an example of a Green Building which suffered from poor performance during the operational stage because of an overly designed CHP system. At the time of the project development, the CHP system helped the project get the required certification and in theory it saved energy. The CHP costed a lot to run on the natural gas to produce electricity and the building tenants couldn't be forced to use the electricity produced. The only places

using that energy were the communal spaces in the building with few lighting fixtures. The more a CHP system is run, the more value it adds, but in the case of this project, it couldn't be used much because of low end-use which points towards the failure of delivering what was expected in terms of operational emissions. This failure was because of poor feasibility study and the wrong expectations from the building occupancy and operation.

3.3 Timeliness of project activities

The conditions within this *theme,* as shown in Figure 3.2, are about the time sensitivity of different aspects of GB project development. The two highly identified conditions within this *theme* are 'Early engagement of project team' (*n* = 30) and 'Early introduction of project targets' (*n* = 22).

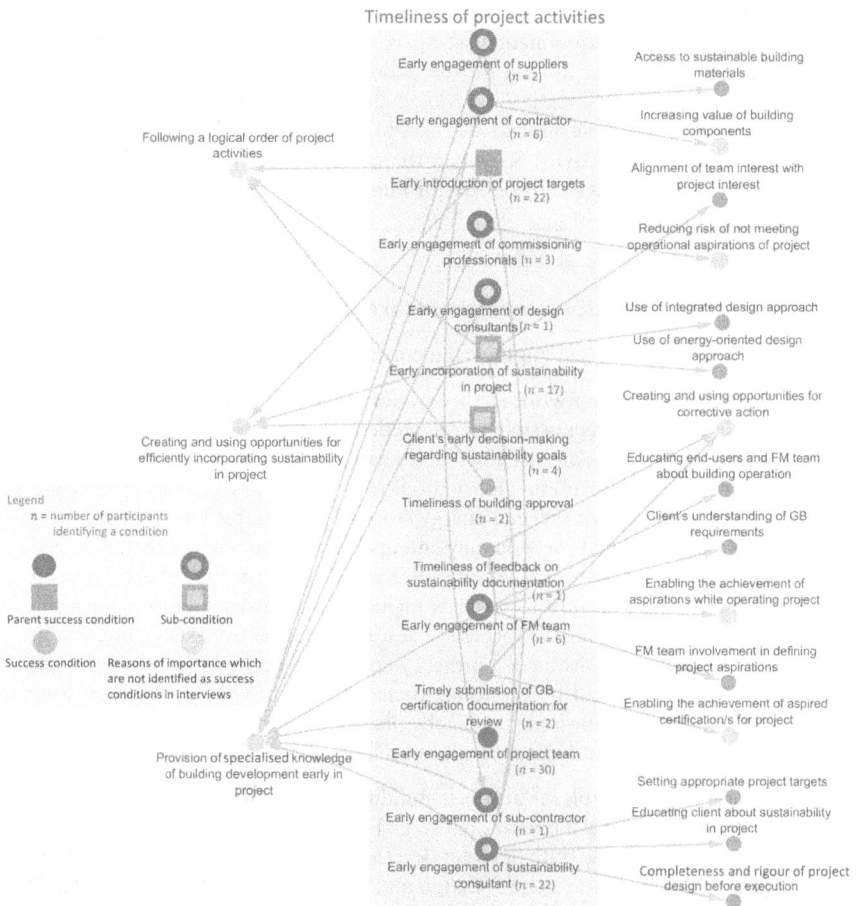

Figure 3.2 Conditions of 'Timeliness of project activities'.

Conditions within this *theme* are specifically related to the project client, contractor, design team, facility management team, sub-contractors, suppliers, sustainability consultant, commissioning team, building control authorities, and GB certification organisations. These conditions are generally related to the overall project team (Table 5.1 and Figure 5.2).

According to the conditions represented by this theme, the likelihood of GB project success increases in case: there is early involvement of team members in the project lifecycle; the project targets are introduced at the project outset; building planning approvals are processed in due time; and the GB certification documentation is timely submitted for review and receives a timely feedback from the GB certification organisation. The early involvement of team members implies early involvement of design consultants, sustainability consultants, commissioning professionals, suppliers, and sub-contractors; contractor's engagement in the design development stage; and the engagement of facility management (FM) team during the project design development. The introduction of project targets at the project outset implies the client's early decision-making regarding sustainability goals; consideration of sustainability from project start; involvement of GB concepts in the early design decisions; and the clear articulation of the idea of GB development in the site selection stage.

These conditions are important for project success as they help set appropriate project goals; enable the involvement of FM team in defining project aspirations; ease the access to sustainable materials; increase the value of building components; align team interests with project goals; and help educate the project client, end-users, and FM team about sustainable building development and operation. These conditions contribute towards design development by ensuring the completeness of design before execution and ensuring the use of integrated and energy-oriented design approach. These conditions are also important for GB project success since they create the opportunities for corrective actions; reduce the risk of not meeting operational aspirations; and in fact enable the achievement of aspired certifications and operational goals. Many of these conditions help follow a logical order of project activities; create opportunities for efficient incorporation of sustainability; and enable the provision of specialised knowledge of building development early in the project.

Conditions within this theme, as shown in Figure 3.2, can reduce the non-value-adding activities in a GB project by enabling the access to sustainable building materials, enabling the alignment of team interest with project interest, contributing towards the client's understanding of GB requirements, ensuring completeness and rigour of project design before execution, following a logical order of project activities, enabling the availability of specialised knowledge of building development early in the project, and by creating and using opportunities for corrective action and opportunities for efficiently incorporating sustainability in the project.

These conditions can also help create value for the client by educating end-users and FM team about building operation, enabling the achievement of aspired certification/s for the project, enabling the achievement of aspirations while operating the project, increasing the value of building components, reducing the risk of not meeting operational aspirations of the project, and setting appropriate project targets.

Early incorporation of sustainability in project

According to a Singapore-based design consultant (SN-M-5),

> If you don't bring in Green within the project right from the start, you will probably never see it.

According to a Singapore-based energy manager (SN-M-12),

> A problem in many Singapore-based projects is that the clients don't engage the sustainability consultants in the planning stage. Some clients only pay attention to sustainability at detailed design stage and engage us at that time or even afterwards. This becomes a challenge for consultants to incorporate the sustainability features in a project. When features are incorporated in the late design stages, it would usually result in extra project costs, which may cause some features to be dropped off. This late engagement practice is a challenge for us to deliver successful Green Building project in the end.

According to a Hong Kong-based sustainability consultant (HK-M-1),

> It's difficult these days for projects to fail in terms of achieving sustainability aspirations. If projects still fail, then it is because of the poor planning. By poor planning it means that the green is incorporated in the middle or at the end of the project design stage and even in the construction stage. Because of this the projects struggle. However, if you plan the projects from the day one, there won't be any challenge.

According to a Pakistan-based sustainability consultant (PK-M-1),

> Often conflicts in the interests arise when Green is not considered early in the project planning. For instance, in case a sustainability consultant is brought on team later in the project development, the sustainability consultant would ask the MEP consultants to follow some standards and approaches that can help with the Green Building certification. Both the MEP consultants and the clients at this point may disagree from the suggestions put forward; the MEP consultants would argue that the additional requirements are not part of the job they signed up for; and the client would argue against the extra spending required for achieving sustainability targets.

Timely submission of GB certification documentation for review

While talking about an Australia-based project, a sustainability consultant (AU-F-3) mentioned,

> When the project development started it was realised that the Green Star rating tool-related requirements were not clear and specific for the project. When the project reached the construction stage, relevant documentation for Green Star assessment was not prepared on time. When the GBCA finally reviewed the project and provided the feedback it was too late. The project was already close to its completion and the aspects which could have been easily incorporated at the design stage were enormously difficult to consider then. Because of the arrival of relevant information late in the project, design changes were difficult to make, and the project couldn't achieve its sustainability aspirations.

3.4 Defining project goals

The approaches used in defining GB project goals are associated with project success. Ten conditions occur in this *theme* (Figure 3.3) among which the key ones are 'End-users' involvement in defining project aspirations' ($n = 8$), 'Project team involvement in defining project aspirations' ($n = 3$), 'Setting appropriate project targets' ($n = 21$), and 'Stringency level of project sustainability requirements' ($n = 4$). Conditions within this *theme* are related to the project client, end-users, FM team, and project management (PM) team, as well as the overall project team (Table 5.1 and Figure 5.2).

According to the conditions represented by this theme, the likelihood of GB project success increases in case the end-users and particularly the facility managers are involved in project decision-making and defining project aspirations; the project requirements are decided in consensus with the project team as the client consults with experts to define the project scope; the project sustainability requirements and aspired certification requirements are not so stringent that they cannot be achieved using the available resources; and the project targets are appropriately set. Appropriately defining project targets implies that project vision is rigorously developed, outcomes are clearly defined, and the project has a well-defined scope; the client provides specific details of the expected building performance; sustainability targets are coherent with the client's aspirations; the project targets mutually benefit all key stakeholders; there is not an overreliance on GB certification systems and point pursuing mentality; sustainability brief is aligned with project budget; and those certification systems are used, which acknowledge the sustainability and innovation of project features.

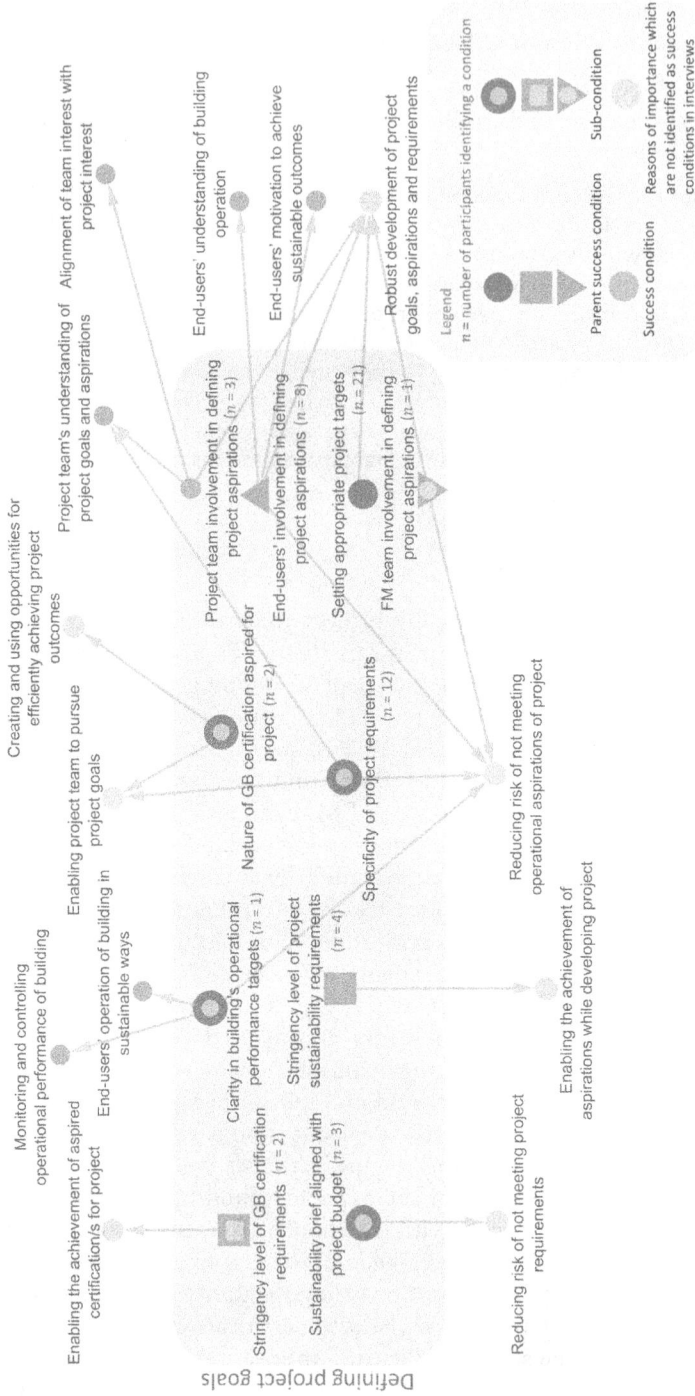

Figure 3.3 Conditions of 'Defining project goals'.

The importance of these conditions for project success is that they help define the appropriate and non-ambiguous targets during early project stages by using the sustainability-related understanding and experience of the project team; help rigorously achieve the sustainability aspirations and ensure that sustainability goals are consistently followed during project development; ensure that the project outcomes are not affected by the limitations of GB certification systems; motivate the project team to deliver project aspirations; and make it easier for the project team to fulfil project goals. These conditions are also important as they ensure that project goals including the operational outcomes are aligned with the client's requirements; enable the end-users to have an understanding of sustainable building operation and take steps to achieve the operational targets; and ensure that the key stakeholders are satisfied from project development.

Conditions within this theme, as shown in Figure 3.3, can reduce the non-value-adding activities in a GB project by creating opportunities for efficiently achieving project outcomes, enabling the project team to pursue project goals, ensuring the team's understanding of project goals and aspirations, and aligning team interest with project interest. These conditions can also help create value for the client by ensuring a robust development of project goals, aspirations, and requirements; resulting in end-users' understanding and motivation of sustainably operating the building; enabling the achievement of aspirations such as GB certifications while developing the project; and reducing the risk of not meeting project requirements and operational aspirations. The condition 'Sustainability brief aligned with project budget' is also related to the transformation of the project.

Specificity of project requirements

According to a Singapore-based engineering consultant (SN-M-6),

> Failure in building performance is sometimes because of the unclear instructions from the building owner. For instance, the owner may say that he wants a certain green rating but in case it is not mentioned in the contract, and it is not mentioned in the meetings, then it would lead to ambiguities and the project team will not know what to achieve.

3.5 Inspection, monitoring, and control

The inspection, monitoring, and control of project development and operation are important aspects associated with GB project success. This *theme* has seven conditions (Figure 3.4) among which the key ones are 'Monitoring of project development' ($n = 12$), 'Inspection of project upon construction' ($n = 8$), 'Monitoring and controlling operational performance of building' ($n = 5$), 'Thoroughness of value engineering exercise' ($n = 3$), and 'Review

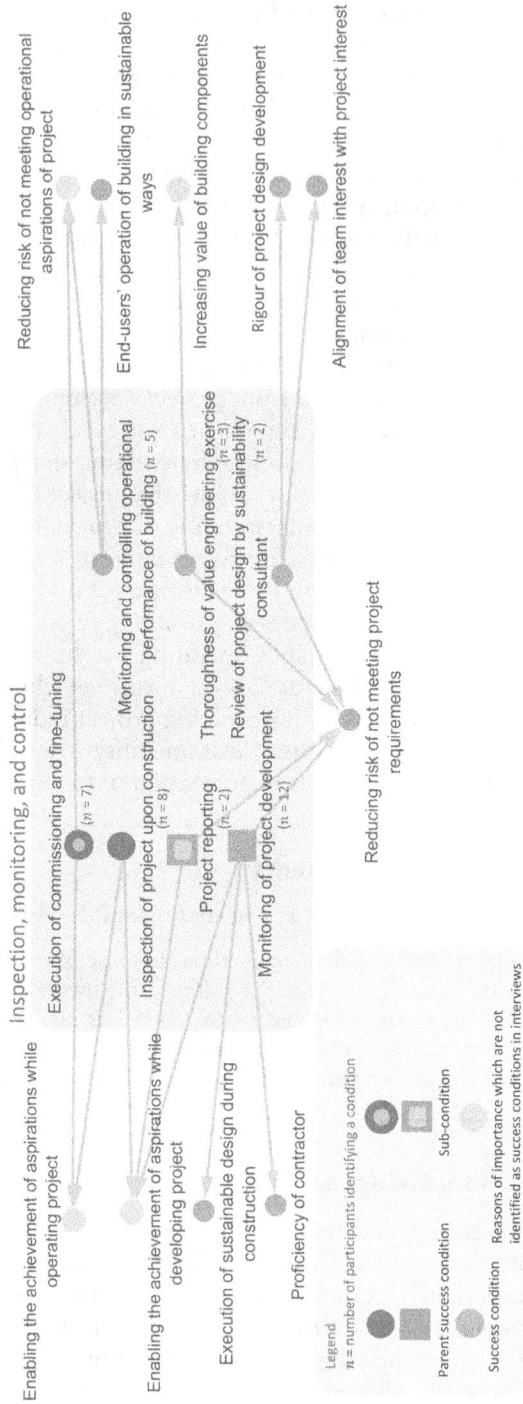

Figure 3.4 Conditions of 'Inspection, monitoring, and control'.

of project design by sustainability consultant' (n = 2). 'Inspection of project upon construction' is a condition representing a *cluster* with one sub-condition which is 'Execution of commissioning and fine-tuning' (n = 7). 'Monitoring of project development' (n = 12) is also a condition representing a *cluster* with one sub-condition which is 'Project reporting' (n = 2). Conditions within this *theme* are related to sustainability consultant, commissioning team, PM team, and FM team, as well as the overall project team (Table 5.1 and Figure 5.2).

According to the conditions represented by this theme, the likelihood of GB project success increases in case the project is monitored during development and inspected upon construction; the sustainability consultant can critically review project design and other aspects of project development and can demand revisions in case the project design is inadequate; value engineering exercises are rigorously conducted to reduce the risk of eliminating sustainable features; the FM team monitors the building performance; a follow-through is conducted to ensure that the users understand the sustainable operation of the building; and regular audits are conducted to report the operational performance of the building. Project inspection upon construction implies that upon completion, the project is thoroughly inspected for conformance with specifications; commissioning, fine-tuning, and seasonal testing of the building are rigorously conducted; and building systems are commissioned by a third party. Monitoring of project development implies that contract administration, quality assurance, and quality control are rigorously performed; quality of design work is managed; building construction is monitored for conformance with specifications; materials used in construction are reviewed; the site is regularly visited by the design consultant; the deliverables for different project stages are kept in account; checkpoints are used for reviewing sustainable development; and project reporting is conducted at regular intervals to see if project requirements are being met, and this reporting is continued throughout the different project stages.

The importance of these conditions for project success is that they ensure that the project team is held accountable for project deliverables and takes interest in the sustainable development of the project; project development is up to the expectations; sustainability aspirations are not compromised during project development; and the building in its operational life performs according to the design intent. In specific terms, these conditions are important for project success since they ensure that design and specifications during construction are effectively executed; project aspirations are met in construction; the design team fulfils sustainability objectives during project development; the shortcomings in design are compensated; the end-users and the project client know if the building in its operational life is performing as intended by design; and the end-users understand the sustainable operation of the building.

Conditions within this theme, as shown in Figure 3.4, can reduce the non-value-adding activities in a GB project by resulting in a rigorous design development; alignment of team interest with project interest; and execution of sustainable design during construction. These conditions can also help create value for the client by enabling the achievement of aspirations while developing and operating the project; increasing the value of building components; enabling sustainable operation of the building by end-users; and reducing the risk of not meeting project requirements and operational aspirations of the project. One condition, that is 'Thoroughness of value engineering exercise', is also related to the transformation of the project.

Execution of commissioning and fine-tuning

While talking about an Australia-based project, a sustainability consultant (AU-M-10) mentioned,

> We wanted to get the credit point for occupant thermal comfort and therefore we specified an under-floor air distribution system for the project. However, during the commissioning phase, the system did not work mainly because of the poor commissioning. A lot of time was spent afterwards to get the system to work efficiently. Another reason for the underperformance was that the contractor was not aware of the particular air conditioning system being used.

3.6 Flow of project information

The flow of project-related information is an important aspect associated with GB project success. Among the six conditions occurring in this *theme* (Figure 3.5), the key ones are 'Clarity in communication of project goals' ($n = 7$), 'Communication among project team' ($n = 11$), 'Project team's access to robust information' ($n = 5$), and 'Smooth transition of project from inception to operation stage' ($n = 5$). Conditions within this *theme* are related to project team and FM team (Table 5.1 and Figure 5.2).

According to the conditions represented by this theme, the likelihood of GB project success increases in case project goals are clearly communicated to project team early in the project, and the client discusses the sustainability brief with the design consultant and contractor and emphasises its importance; project team members communicate effectively and have the access to robust information; smooth transition of project occurs from inception to operation stage by using approaches such as Soft Landings; and there is a smooth handing over of the project to the contractor once the tender is awarded. Team access to robust information implies that the team has timely access to information regarding different aspects of project development including building materials,

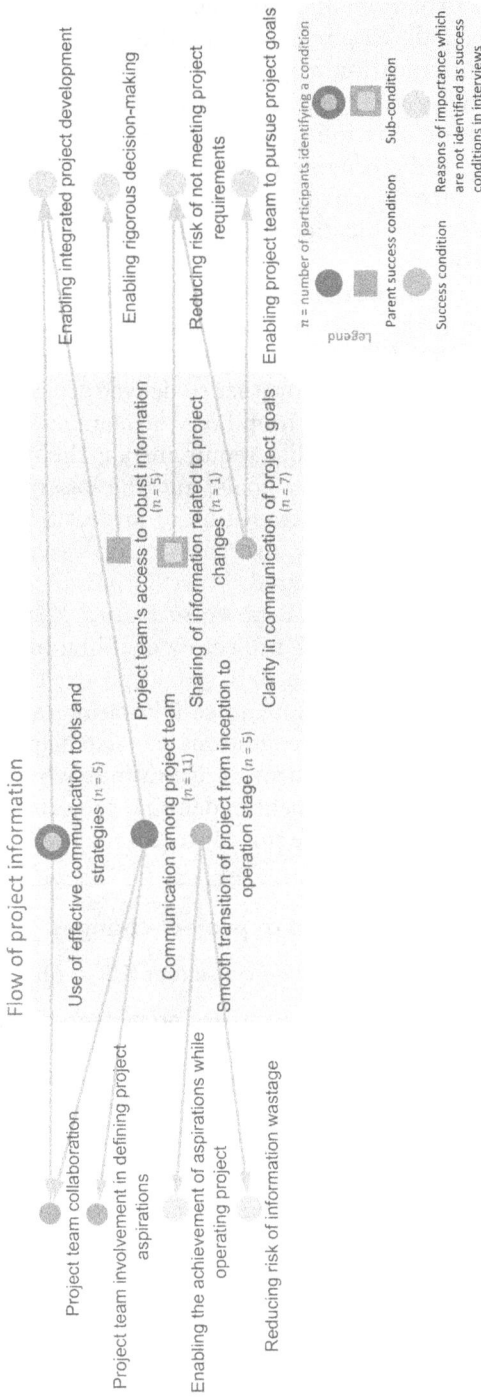

Figure 3.5 Conditions of 'Flow of project information'.

and the project information and specifically the information related to changes in project development is shared among the project team. Effective team communication implies that the project team members communicate effectively and clearly to work collaboratively and resolve project issues; the multi-disciplinary design team has meetings to discuss green initiatives and green implementation; workshops of the project team are conducted on a regular basis from inception to completion; and an open and honest communication process is established with the design team early in the project.

These conditions are important for project success as they ensure that the project team can make rigorous decisions; the project team members are aware of the project deliverables, know how their work is related to the work of others, and can work together to deliver project outcomes; the project team can provide an opinion on how the changes in project development may affect the sustainability requirements; the design team can contribute in defining sustainability goals; and the loss of project-related information is avoided during the transition of project from one party to another, and in this transition, the awareness of end-users and FM team regarding building operation is ensured.

Conditions within this theme, as shown in Figure 3.5, can reduce the non-value-adding activities in a GB project by enabling integrated project development; empowering project team to pursue project goals; supporting rigorous decision-making; enabling team collaboration; and reducing the risk of information wastage. These conditions can also help create value for the client by enabling the achievement of aspirations while operating the project; ensuring the team involvement in defining project aspirations; and reducing the risk of not meeting the project requirements.

Sharing of information related to project changes

According to a UK-based sustainability consultant (UK-F-6),

> The project changes if not communicated properly to the sustainability people become one of the biggest reasons of underperformance in Green Building projects. Sometimes when changes are made, some aspects of the changes on the sustainability performance are checked but some aspects are missed. A lack of focus on the details affects sustainable performance.

> Although the Value Engineering exercise is developed to involve all the project team members and require the consent of everyone involved while making changes, sometimes details of the changes being made in project design get missed and some of the sustainable attributes of the project are compromised. This is because construction and engineering are huge industries involving a lot of documentation and the challenge is to connect the information with relevant people in a project.

Project team's access to robust information

According to a Singapore-based sustainability manager (SN-M-2),

> Sometimes trouble happens when the project information is not communicated rightly. In case of one of our projects, the drawings mentioned a pantry. This pantry was actually a 150-seat restaurant. If you have such distortions in information, then it becomes hard to plan how the building would save energy and achieve its aspired certification. So, in case of that project, we had to shift down from the Platinum level to the Gold level.

Smooth transition of project from inception to operation stage

According to a Singapore-based facilities manager (SN-M-3),

> Having BIM in construction will lead to better flow of information, and lesser arguments in project delivery. In FM sector these days we have BIM-FM. In this approach, the construction models are transformed into the usable models for building operation.

3.7 Planning approach

The project planning aspects are associated with GB project success. This *theme* has six conditions (Figure 3.6) among which the key ones include 'Rigour of project planning' (*n* = 19), 'Attention towards details' (*n* = 5), and 'Rigour of risk management' (*n* = 4). 'Rigour of project planning' is a condition representing a *cluster* with three sub-conditions which are 'Adequate budget allocation for project development' (*n* = 13), 'Adequate time allocation for project development' (*n* = 11), and 'Adequate time allocation for commissioning and testing' (*n* = 1). Conditions within this *theme* are related to the project team and mainly to the PM team (Table 5.1 and Figure 5.2).

According to the conditions represented by this theme, the likelihood of GB project success increases in case the project is rigorously planned to consider all the aspects of development; attention is given to detailing, finishing, and workmanship in the project; attention is given to key details in project development and operation and even the small details (for example, instrumentation and measurement of the relative temperatures); and rigorous risk management is performed by which risks are regularly assessed, remediation actions are promptly taken, and risks are managed in terms of the project objectives. Rigour of project planning implies that there is an appropriate budget allocation to support the sustainability initiatives in project development; the project budget aligns with project scope; buffers

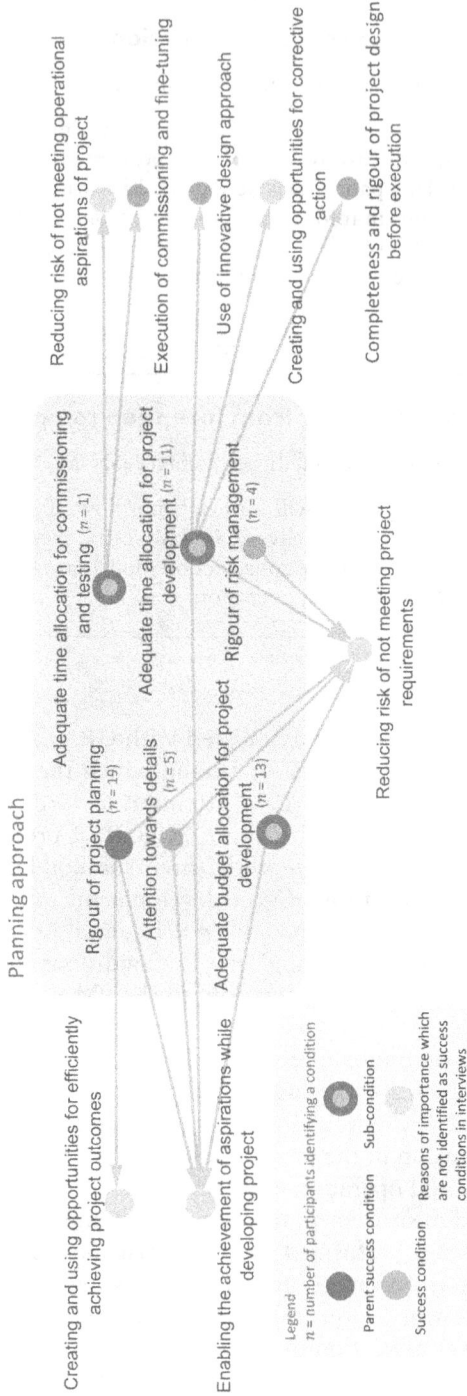

Figure 3.6 Conditions of 'Planning approach'.

are provided in budget; adequate time is allocated for project stages such as inception, planning, and design and particularly the commissioning and testing exercise; and the time allocation considers sustainability aspects in project development.

These conditions are important for project success as they enable rigour and accuracy in project development; ensure a preparedness while the project is being developed; enable the fulfilment of aspirations while developing the project; reduce the risk of not meeting project requirements; help ensure the completeness of design deliverables before execution; and avoid the lack of innovative design approaches resulting from time shortage. These conditions are also important as they help avoid the risk of inappropriately conducted commissioning and testing resulting from a time shortage, and the risk of eliminating project features resulting from the retrospective realisation that the allocated time or budget is insufficient.

Conditions within this theme, as shown in Figure 3.6, can reduce the non-value-adding activities in a GB project by resulting in a complete and rigorously developed project design before the start of construction, and creating opportunities for efficiently achieving project outcomes. These conditions can also help create value for the client by enabling the achievement of aspirations while developing the project; creating opportunities for corrective action; and reducing the risk of not meeting project requirements and operational aspirations. Two conditions which are 'Adequate budget allocation for project development' and 'Rigour of project planning' are also related to the transformation of the project.

3.8 Completeness and rigour of deliverables

The completeness and rigour of deliverables is an important aspect associated with GB project success. The three conditions occurring in this *theme* (Figure 3.7) are 'Setting of a detailed sustainability charter or brief' (n = 10), 'Completeness and rigour of project design before execution' (n = 9), and 'Completeness of project documentation for certification' (n = 2). Conditions within this *theme* are related to the client, contractor team, design team, and the overall project team (Table 5.1 and Figure 5.2).

According to the conditions represented by this theme, the likelihood of GB project success increases in case a detailed sustainability charter or brief is developed, and the project brief is based on the needs of project stakeholders; the project design is complete and well defined before the start of execution particularly in terms of sustainability initiatives; and the documentation for third-party GB certification is complete. The importance of these conditions for project success is that they contribute to the project team's understanding of sustainability objectives; enable early incorporation of sustainability in the project development; ensure the suitability of the project design for execution; reduce the possibility

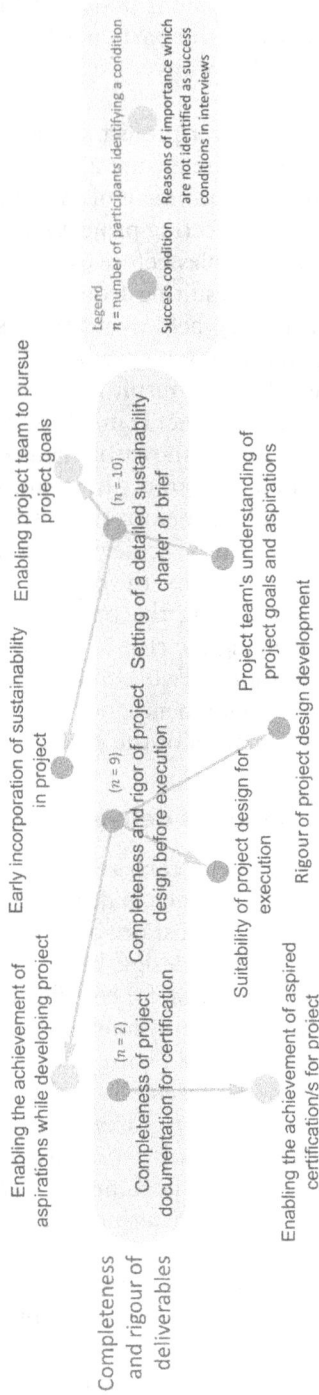

Figure 3.7 Conditions of 'Completeness and rigour of deliverables'.

of design-related problems during the construction stage; and enable the achievement of aspired sustainability certification and the effective achievement of project goals.

The conditions related to this *theme,* as shown in Figure 3.7, are about the completeness and rigour of project deliverables. Some of these conditions reduce the non-value-adding activities in a GB project by ensuring the suitability of the project design for execution, enabling a rigorous design development, ensuring team's understanding of goals, and contributing towards the early incorporation of sustainability. Conditions within this *theme* also help create value for the client by enabling the achievement of aspirations such as GB certifications and enabling the project team to pursue project goals.

Completeness and rigour of project design before execution

According to a UK-based engineering consultant (UK-M-4),

> In case of refurbishment projects, sometimes building services engineers are given unrealistic deliverable timelines from the clients. In these cases the deliverables from the consultants may not be complete and a contractor might need to use his own expertise to bridge the knowledge gaps.

Completeness of project documentation for certification

According to an Australia-based sustainability consultant (AU-F-5),

> Green Building projects may not be able to meet the aspired sustainability levels when they fail to meet the documentation requirements. This can happen as Green Star certification requires an immense amount of documentation which may not be possible in case the people in the project team particularly the consultants and contractors are beyond approach. This may also happen in case no one on the project is interested in collecting this documentation. This problem can significantly result from the head contractor's side, who is responsible for collecting as much as ninety percent of the documentation. Before the building gets certified, the contractor may already have finished his job on the project and left for another project. Upon finishing of the project, it may also become difficult to collect documentation from the smaller players like subcontractors as they may have already left or gone out of business.

Setting of a detailed sustainability charter or brief

According to a UK-based sustainability consultant (UK-F-6),

> Sustainability needs to be made a part of the project agenda to give it a
> timely focus.

3.9 Changes during project development and fulfilling design intent

Changes occurring during project development affect GB project success, and often, when the development follows the plan, it usually results in GB project success. These aspects are considered together in this *theme* since the same conditions account for both. Some key conditions occurring in this *theme* include the following: 'Change in project team members' ($n = 2$), 'Execution of sustainable design during construction' ($n\ 5$), and 'Scope changes during project execution' ($n = 9$). Conditions within this *theme* are related to the project clients, contractors, and in some cases the overall project team (Table 5.1 and Figure 5.2). While Figure 3.8 shows the conditions and sub-conditions within this theme, it also depicts the reasons why these conditions are important for project success. Similar figures are used for the rest of 20 themes to present conditions and show the reasons why these conditions are important for project success.

According to the conditions represented by this theme, the likelihood of GB project success increases in case the project team remains unchanged during the project development and project construction is as per design specifications. GB project success is also enabled when the scope changes are reduced during project execution, the client's requirements and the project design remain unchanged, and the client is motivated to achieve the original project intent. These conditions as shown in Figure 3.8 are important for project success as they facilitate team synergies, reduce the gap between design speculations and reality, reduce the risk of losing project information, enable the project team to pursue goals, ensure the completeness of project documentation for GB certification purpose, and ensure the achievement of aspirations in project development.

Conditions within this *theme* can reduce the non-value-adding activities in a GB project by reducing the risk of information wastage and by establishing team synergies. These conditions can also help create value for the project client by helping achieve the aspirations during the project development, enabling the project team to pursue project goals, and ensuring the completion of project documentation for aspired certification. The condition regarding the 'execution of sustainable design during construction' is also related to the transformation of the project.

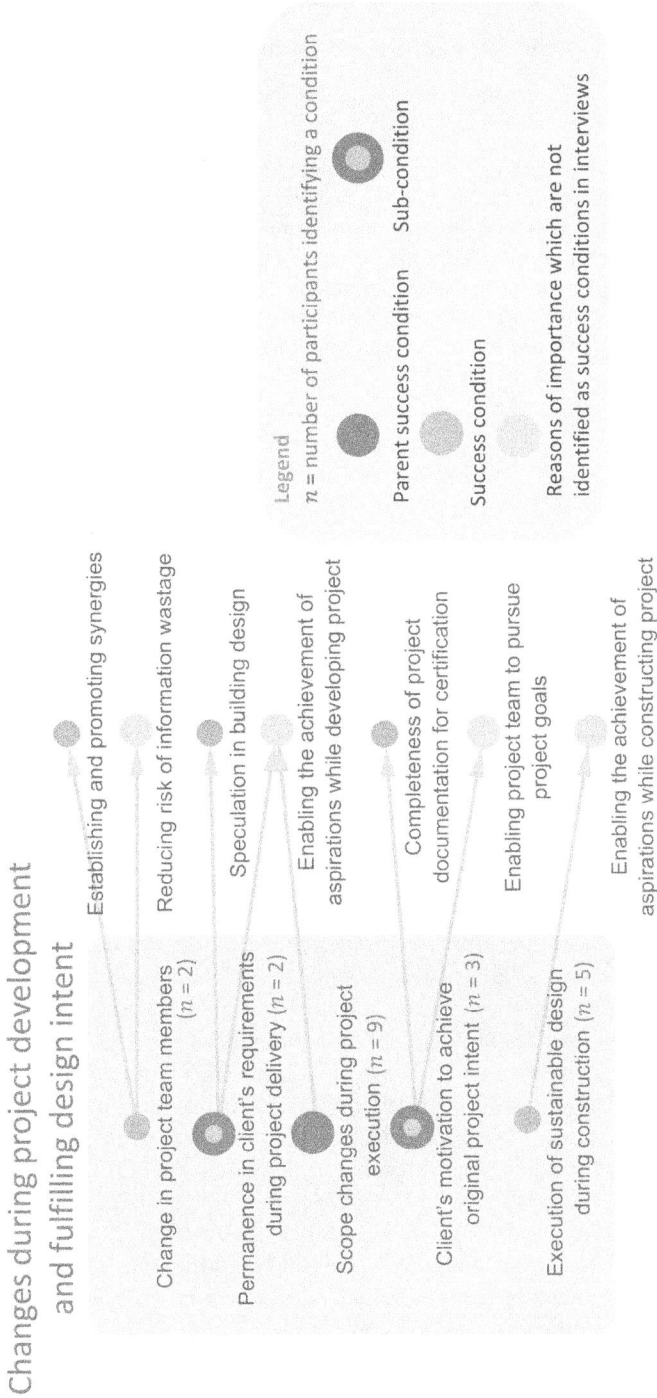

Figure 3.8 Conditions of 'Changes during project development and fulfilling design intent'.

Permanence in client's requirements during project delivery

While talking about an office building, a UK-based sustainability consultant (UK-F-3) mentioned,

> The project is a London based office block. A good project design was proposed earlier in which the building was fully naturally ventilated. Just before the submission of the planning documentation, the operational team of the building decided that they wanted Air Conditioning in the meeting rooms. However, the building then had to be fully air conditioned for satisfying everyone and this way the plan to naturally ventilate the building was ruined. This is a fairly common thing in projects since the client's requirements change during the project delivery.

Client's motivation to achieve original project intent

According to a Hong Kong-based sustainability consultant (HK-M-2),

> A major reason that projects cannot achieve their targets is that the client gives up. For instance, in one of my projects the certification requirement was to provide 40% area for greenery. The client had to trade-off between the greenery area and the building-related facilities. However, finally the client had to give up and provided only 20% area for greenery. As a result, the project had to go for a lower rating and not the one which it originally aspired.

3.10 Clarity in project development

Clarity in the project development process is an important aspect associated with GB project success. This *theme* (see Figure 3.9) has six conditions among which the key ones are the 'Clarity in process of developing project' ($n = 9$) and 'Delegating clear responsibilities to project team' ($n = 6$). The 'Clarity in process of developing project' is a condition representing a sub-condition, which is the 'Use of clearly defined and standardised approaches for GB development' ($n = 5$). 'Delegating clear responsibilities to project team' is a condition representing three sub-conditions, which are 'Project team contractually required to deliver sustainable outcomes' ($n = 2$), 'Provision of sustainability specifications and other related information in tender' ($n = 4$), and 'Specificity of deliverables from Design Consultants' ($n = 2$). Conditions within this *theme* are related to the client, design team, and contractor team, as well as the overall project team (Table 5.1 and Figure 5.2).

According to the conditions represented by this theme, the likelihood of GB project success increases in case clear responsibilities are delegated to

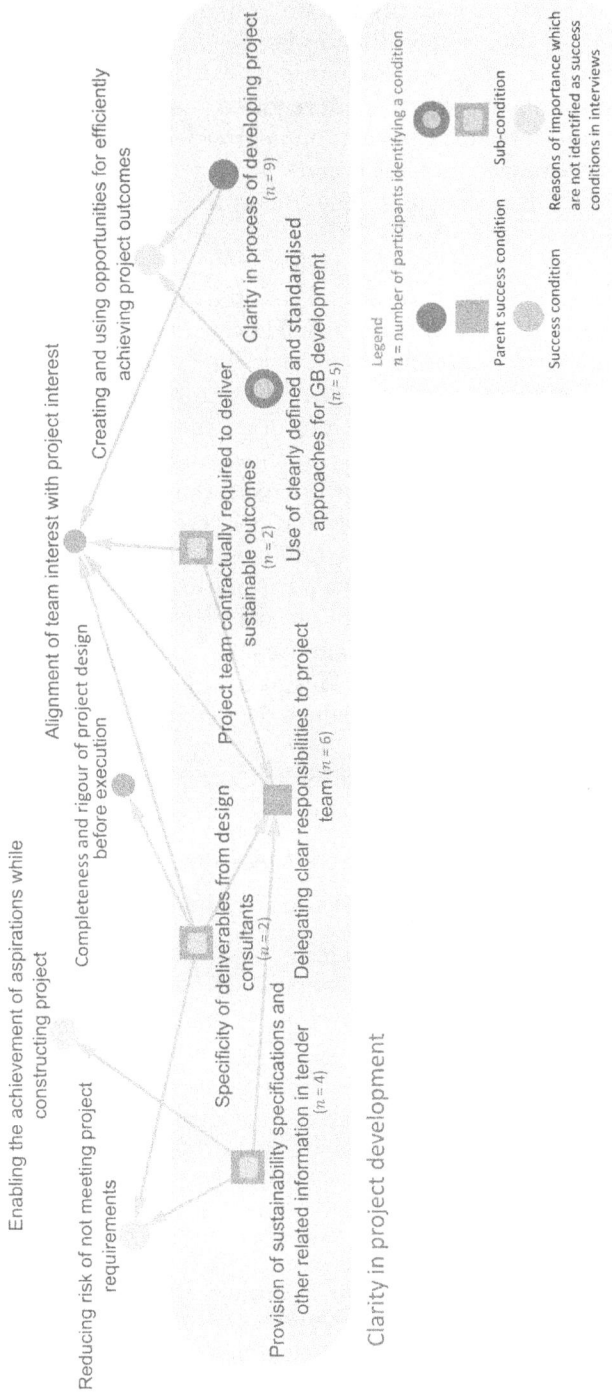

Figure 3.9 Conditions of 'Clarity in project development'.

the project team and there is clarity and transparency in the project development process. Delegating responsibilities implies that the project team, and particularly the contractor team, is contractually required to deliver sustainable outcomes; specific deliverables are required from the design team; and the necessary project-related information is provided in tender documents. Clarity in the development process implies that a project development strategy exists from the project outset. The importance of these conditions for project success is that they ensure a team focus on project goals and align team interests with project goals; enable the project team to accomplish project goals in an effective way; ensure that the project team, particularly the contractor and the design team, is held accountable for what they are required to achieve in a project; ensure that the focus on sustainability is maintained during project development; and reduce the possibility of incomplete project design outcomes resulting from unclear deliverables.

The conditions related to this *theme,* as shown in Figure 3.9, ensure a rigorous and complete project design before execution, enable the alignment of team interest with project interest, and help create and use the opportunities for efficiently achieving project outcomes. Owing to these reasons, the conditions within this *theme* reduce the non-value-adding activities in a GB project. Moreover, these conditions also result in reducing the risk of not meeting project requirements and enable the achievement of aspirations during project construction. This means that these conditions also help create value for the client and reduce the gap between potential value and actual value by enabling the achievement of aspirations during project development. The condition 'Delegating clear responsibilities to project team' is also related to the transformation of the project.

Provision of sustainability specifications and other related information in tender

While talking about an office fit-out project, a UK-based sustainability consultant (UK-F-1) mentioned,

> We reviewed the project documentation against the project requirements at the time of tendering. We provided a detailed feedback at that time that the project tender was not meeting the project requirements and that the design had to be changed in some areas in order to meet the project requirements. We were assured that the design will be changed, but later the issues were ignored. This resulted in project failure .... Although we as sustainability team developed a document to go with tender, we were not invited to meet the contractor while the contractor was brought onboard. The contractor was not aware of many basic things related to sustainability and in case we had the opportunity of meeting the contractor earlier, those mistakes might have been avoided.

Clarity in the process of developing a project

According to a UK-based sustainability consultant (UK-M-2),

> Unless the project aspirations are very clear and the process of developing the project is well-defined, a Green Building project is moving towards failure. The failure may not happen in case of some exceptions; however, it will happen in case of 90% of the projects because of these reasons.

3.11 Engraving sustainability in project development

Engraving sustainability in project development is an important aspect associated with GB project success. The three conditions occurring in this *theme* (Figure 3.10) are 'Engraving sustainability in project development' ($n = 8$), 'Procurement of project site based on sustainability goals' ($n = 3$), and 'Use of environmental management systems in construction' ($n = 1$). Conditions within this *theme* are related to the project client, design team, and contractor team (Table 5.1 and Figure 5.2).

According to the conditions represented by this theme, the likelihood of GB project success increases in case sustainability is engraved and thoroughly considered in different aspects of project development instead of being considered in a bolt-on approach; well-developed environmental management systems are used in construction stage; and the procurement of the project site is in accordance with the sustainability aspirations of the project. The importance of these conditions for project success, as shown in Figure 3.10, is that they result in increased project value and contribute towards the achievement of project sustainability goals during project development.

The conditions related to this *theme* are about engraving sustainability in different aspects of project development. These conditions reduce the non-value-adding activities in a GB project by enabling an integrated design approach and by creating and using the opportunities for efficiently achieving project outcomes. Conditions within this *theme* also help create value for the client by enabling the achievement of aspirations while developing the project.

Engraving sustainability in project development

According to a UK-based sustainability consultant (UK-M-2),

> Sustainability should be a part of the project development and delivery process. Sometimes, it's just considered a project requirement, implemented in a bolt-on way, and is not seeded in the process. When sustainability is added as a bolt-on, it ruins the whole process and the efficiency of the final product.

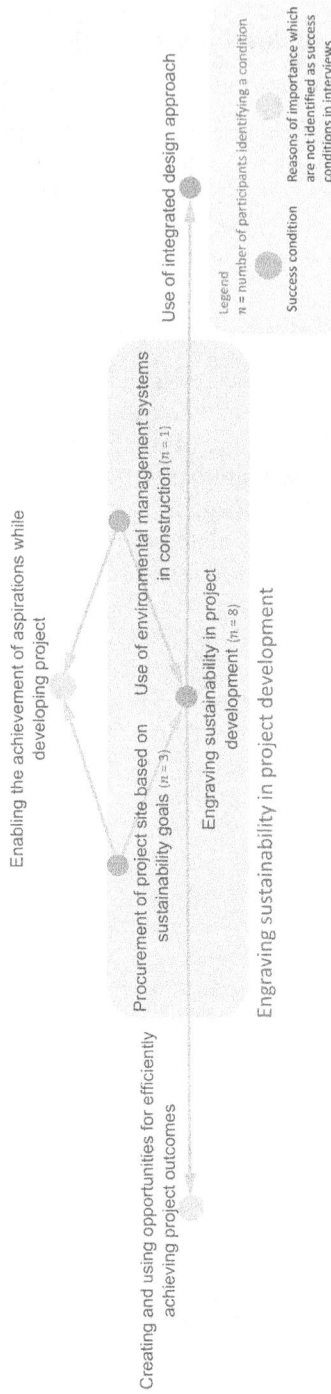

Creating and using opportunities for efficiently achieving project outcomes

Enabling the achievement of aspirations while developing project

Procurement of project site based on sustainability goals (n = 3)

Use of environmental management systems in construction (n = 1)

Engraving sustainability in project development (n = 8)

Use of integrated design approach

Engraving sustainability in project development

Legend
n = number of participants identifying a condition

Success condition

Reasons of importance which are not identified as success conditions in interviews

Figure 3.10 Conditions of 'Engraving sustainability in project development'.

3.12 Constraints

The project constraints are important aspects associated with GB project success. The three conditions occurring in this *theme* (Figure 3.11) are 'Accessibility of project funding' ($n = 6$), 'Access to sustainable building materials' ($n = 2$), and 'Ease of logistics at project location' ($n = 1$). Conditions within this *theme* are related to the client, end-users, design team, and suppliers (Table 5.1 and Figure 5.2). According to the conditions represented by this theme, the GB project performance is affected by the access to sustainable building materials, availability of funds to meet the project goals, and the accessibility of project location from a logistics viewpoint. The importance of these conditions for project success is that they enable the achievement of aspirations while developing the project.

The conditions within this *theme* act as constraints in GB project development, and these conditions affect the value for the client as they are related to the achievement of aspirations during project development. These conditions are also related to the transformation of the project.

Accessibility of project funding

According to a UK-based design consultant (UK-F-2),

> Financial problems hurt a project. Upon realizing that the set budget is not sufficient, some of the project features are stripped out through the value engineering process and normally it is the efficient features of a project which are first eliminated in the value engineering exercise.

According to a Hong Kong-based sustainability consultant (HK-M-1),

> Getting a silver or a gold certification will be straight forward unless you don't have enough budget. The costliest items are the good IAQ as you need to have good air-conditioning systems. If you can afford energy efficient equipment, achieving certifications for the project won't be difficult for you.

3.13 Summary

This chapter is the second among the two chapters providing a detailed account of GB success conditions identified by 75 industry experts from Australia, the UK, the UAE, Singapore, Hong Kong, and Pakistan. Unlike Chapter 3 related to the social conditions of project success, this chapter addressed the technical success conditions of GB projects. Together, these attributes constitute 11 broader themes comprised of 38 success conditions and 43 sub-conditions. This chapter discussed the changes during project development and the means to fulfil design intent; clarity in project development; approaches to define project goals and deliverables; project

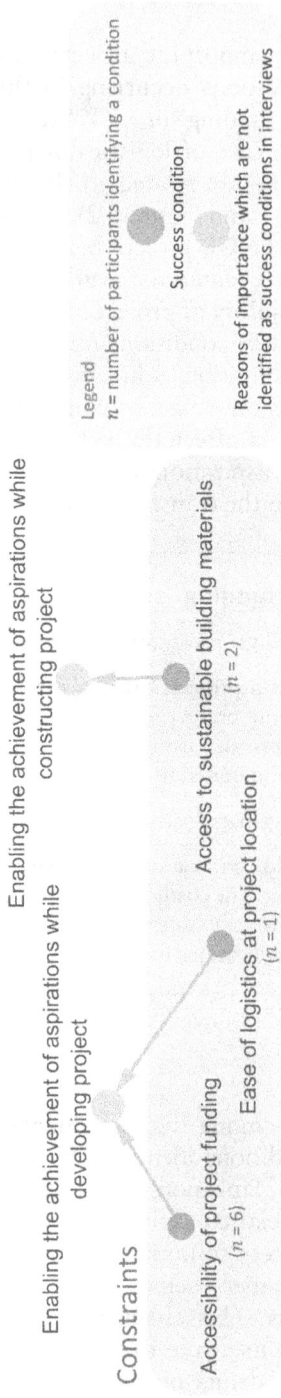

Figure 3.11 Conditions of 'Constraints'.

constraints; design methodology; the process of engraving sustainability in project development; the flow of project information; approaches regarding planning, inspection, monitoring, and control of the project; and timeliness of project activities. In interviews, GB experts have mostly emphasised towards design methodology (n = 41), timeliness of project activities (n = 39), and the process of defining project goals (n = 31).

Chapter 4 will provide a detailed overview of the success conditions in terms of the theoretical interpretation of success conditions and the relevance of success conditions to the credits of GB certification systems.

References

Demaid, A., & Quintas, P. (2006). Knowledge across cultures in the construction industry: Sustainability, innovation and design. *Technovation, 26*(5), 603–610. doi:10.1016/j.technovation.2005.06.003

Gluch, P., Gustafsson, M., & Thuvander, L. (2009). An absorptive capacity model for green innovation and performance in the construction industry. *Construction Management and Economics, 27*(5), 451–464. doi:10.1080/01446190902896645

Hwang, B. G., & Tan, J. S. (2012). Green building project management: Obstacles and solutions for sustainable development. *Sustainable Development, 20*(5), 335–349. doi:10.1002/sd.492

Green Building success conditions

Interpretations, implications, and limitations

4.1 Introduction

> **Who should read this chapter**
>
> Read this chapter, if you are interested in the following:
>
> - Understanding how the GB success conditions associate with Transformation-Flow-Value-generation (TFV) theory
> - Understanding the relevance of success conditions with credits of Green Building (GB) certification systems
> - Understanding the limitations of the findings reported in this study
> - Understanding the ways by which future studies can develop the knowledge on GB success further
>
> If you are interested in the implications of GB success conditions for project stakeholders, see Chapter 5.

To account for the significance of GB success conditions, it is important to look at them from both theoretical and practical lenses, which is a key objective of this chapter. The structure of this chapter is such that first the effect of the interview themes and attributes of interview participants on the identified success conditions is analysed. Subsequently, a comparison of GB success conditions with the success factors of socio-technically complex projects is provided. This is followed by an interpretation of success conditions in terms of the TFV theory. To address the practical implications of research findings, a comparison of the success conditions with the credits of GB certification systems (that is, Living Building Challenge, LEED, BREEAM, and Green Star) is provided. This helps understand how some success conditions are already being implemented by GB certification systems and what are the potential possibilities in this regard. Lastly, the limitations associated with the findings of this study are reported and the potential means for the future research to develop the knowledge on GB project success are discussed.

DOI: 10.1201/9781003322740-6

4.2 Success conditions: Effects of interview participants' attributes

The overall 155 conditions and sub-conditions are classified within 20 broader *themes*. To see how the attributes of interview participants and their viewpoints affect the identification of success conditions within their broader *themes*, the analysis of the frequency distribution of broader *themes* is conducted in this section (shown in Table 4.1).

- **Viewpoints used in interviews:** Regarding the three viewpoints used in interviews for the inquiry of success conditions, it can be observed that varying consideration is given to different *themes* while using the three viewpoints. Success conditions regarding 'Education', 'Completeness and rigour of deliverables', 'Flow of project information', and 'Inspection, monitoring, and control' are mostly identified by participants while using the viewpoint of *Findings from participants' overall experience*, and not by the other two viewpoints. Success conditions regarding 'Constraints' and 'Team authorities, responsibilities, and contractual relationships' are mainly identified by participants while using the viewpoint of *Findings from successful/failed projects*, and not by the other two viewpoints.
- **Regional belonging of interview participants:** Upon analysing success conditions identified by participants with GB experience based in different regions, it can be observed that conditions regarding 'team authorities, responsibilities, and contractual relationships' have received a major focus from the UAE-based participants. Conditions regarding 'changes during project development and fulfilling design intent', 'education', and 'team mindset and priorities' have received no attention from the UAE-based participants. Conditions regarding 'Inspection, monitoring, and control' and 'education' have received insignificant attention from the UK-based participants. Conditions regarding 'planning approach' have received insignificant attention from the Hong Kong-based participants. Conditions regarding 'team commitment to the project' have received insignificant attention from Hong Kong, Singapore, and the UAE-based participants. Conditions related to 'team authorities, responsibilities, and contractual relationships' have received insignificant attention from Hong Kong and Australia-based participants.
- **Experience of participants in GB projects:** Upon analysing success conditions identified by participants with medium, high, and very high experience of working on GBs, it is observed that conditions regarding 'client's characteristics' have received the most attention from participants with medium-level experience of GBs. On the contrary, this group of participants has paid the least attention to conditions regarding

Table 4.1 Distribution of success condition themes in terms of the experience, regional belonging, and viewpoints of participants

Themes of success conditions	Viewpoints			Participant's primary role in GB projects (%)		Experience of participants in GB projects			Country where GB experience of participants is mainly based					
	Findings from a successful/failed projects	Findings from Participants' overall experience	Suggestions for clients	Design consultant	Sustainability consultant	Low to medium experience (less than ten years)	High experience (10 or more years but less than 15 years)	Very high experience (15 years or more)	Pakistan	UAE	Hong Kong	UK	Singapore	Australia
Constraints	7	3	0	16%	5%	13%	13%	11%	50%	0%	0%	15%	7%	17%
Engraving sustainability in project development	0	7	4	16%	10%	8%	9%	22%	0%	0%	0%	23%	7%	17%
Team authorities, responsibilities, and contractual relationships	7	1	1	11%	14%	21%	9%	6%	50%	80%	0%	15%	7%	3%
Clarity in project development	3	12	0	26%	14%	17%	25%	6%	0%	20%	13%	31%	7%	20%
Team procurement methodology	4	8	2	26%	10%	17%	22%	11%	50%	20%	13%	38%	14%	10%
Changes during project development and fulfilling design intent	4	10	1	16%	24%	17%	13%	33%	50%	0%	25%	23%	14%	20%
Education	1	18	0	26%	24%	13%	22%	44%	0%	0%	25%	8%	36%	33%
Team mindset and priorities	7	10	0	32%	19%	17%	22%	39%	0%	0%	38%	54%	14%	20%
Completeness and rigour of deliverables	4	17	2	26%	33%	21%	25%	39%	50%	20%	25%	46%	14%	27%
Client's characteristics	8	13	5	32%	38%	46%	25%	28%	0%	40%	25%	46%	29%	33%
Planning approach	8	17	1	37%	38%	33%	34%	28%	0%	60%	13%	46%	29%	33%
Flow of project information	5	21	1	53%	19%	8%	53%	33%	0%	20%	25%	54%	29%	37%

(Continued)

Table 4.1 Distribution of success condition themes in terms of the experience, regional belonging, and viewpoints of participants *(Continued)*

Themes of success conditions	Viewpoints			Participant's primary role in GB projects (%)		Experience of participants in GB projects			Country where GB experience of participants is mainly based					
	Findings from a successfull/failed projects	Findings from overall experience	Suggestions for clients	Design consultant	Sustainability consultant	Low to medium experience (less than ten years)	High experience (10 or more years but less than 15 years)	Very high experience (15 years or more)	Pakistan	UAE	Hong Kong	UK	Singapore	Australia
Inspection, monitoring, and control	2	22	1	37%	24%	25%	38%	39%	0%	20%	50%	23%	36%	40%
Team commitment to the project	9	20	0	42%	43%	29%	41%	39%	50%	20%	13%	54%	21%	47%
Team collaboration	14	18	0	32%	48%	58%	41%	28%	100%	60%	38%	69%	36%	27%
Defining project goals	14	21	5	53%	38%	38%	47%	39%	0%	60%	63%	62%	50%	27%
Team characteristics	9	29	3	47%	62%	50%	47%	50%	100%	40%	50%	62%	36%	50%
Timeliness of project activities	11	31	10	53%	67%	58%	50%	50%	100%	60%	75%	54%	50%	47%
Design methodology	11	35	3	63%	52%	42%	59%	67%	0%	60%	63%	54%	71%	53%
Cooperation and interest of stakeholders	24	22	6	74%	43%	54%	53%	67%	100%	60%	75%	77%	43%	50%
Number of participants	35	66	19	19	21	24	32	18	2	5	8	13	14	30

'flow of project information' and 'engraving sustainability in project development'. Participants with high-level experience have given much attention to conditions regarding 'flow of project information'. This group of participants has, however, given little attention to 'engraving sustainability in project development'. Compared to participants with medium-level and high-level experience, participants with very high-level experience have given much attention to conditions regarding 'education'. This group of participants has, however, given little attention to 'team collaboration' and 'clarity in project development'.

- **Role of participants in GB projects:** Upon analysing success conditions identified by participants having different professional roles in GBs, it is observed that conditions regarding 'constraints', 'team procurement methodology', and 'flow of project information' have received insignificant attention from sustainability consultants. Conditions regarding 'design methodology', 'flow of project information', and 'cooperation and interest of stakeholders' have received much attention from design consultants. However, these professionals have paid little attention to conditions regarding 'changes during project development and fulfilling design intent' and 'team authorities, responsibilities, and contractual relationships'.

Overall, it can be stated that the identification of success conditions by interview participants is considerably determined by the viewpoints of participants as well as their experience, professional role, and regional belonging. This section has provided analysis at the level of broader *themes* of success conditions. For detailed analysis at the level of success conditions, see Annex B (Section 7.3 and Section 7.4).

4.3 Theoretical implications and interpretations of research findings

This section provides some of the key highlights regarding the interpretation of success conditions in terms of the TFV theory. The theoretical implications of research findings are also explored by comparing GB success conditions with success factors of other socio-technically complex projects.

4.3.1 Comparison of Green Building success conditions with success factors of other socio-technically complex projects

Socio-technical complexity occurs in projects which are about designing, managing, or transforming complex systems. This is particularly relevant to large, innovative, and complex projects, involving many individuals and heterogeneous disciplines. Owing to the sustainability requirements of GBs, these projects, in comparison with traditional building projects, have additional success criteria, more heterogeneous project teams, and a high requirement for technological innovations. This indicates that the development of GB projects has increased socio-technical complexity resulting from the combination of social and technical aspects.

To evaluate if the success conditions of GB projects also occur in case of other socio-technically complex projects, a review of research on the success of complex systems in other disciplines was conducted. While one of the reviewed studies is about construction projects (Duy Nguyen, Ogunlana, & Thi Xuan Lan, 2004), the other reviewed studies were focused on large-scale complex projects (Pisarski et al., 2011), complex multinational projects (Thamhain, 2013), international development capacity-building projects (Ika & Donnelly, 2017), pest management projects (Schmidt, Stiefel, & Hürlimann, 1997), global information systems (Biehl, 2007), and aerospace and defence projects (Rodriguez-Segura, Ortiz-Marcos, Romero, & Tafur-Segura, 2016). In contrast to the term 'Success conditions' used for the findings of this study, the success enablers in reviewed studies are referred to as 'success factors' for the distinguishing purpose.

The comparison shows a considerable representation of GB success conditions in the success factors of other complex systems. Some success factors occurring in reviewed studies represent very specific aspects of a project type and hence do not represent success conditions identified in this study. The representation of success conditions by success factors of complex systems is shown in Table 4.2. The GB success conditions represented by success factors of other complex systems are stakeholder interest; client's characteristics; communication; team characteristics, responsibilities, and commitment; team collaboration; team and client's education; client's funding; project planning, monitoring and control; team selection criteria; defining project; and clarity in the project development process. These success conditions are either about the stakeholder attributes in projects or approaches contributing to successful project development through the efficient use of human skills, knowledge, and commitment. The representation of these GB success conditions by the success factors of other complex systems indicates that the existence of similar success conditions across different complex systems can be due to the socio-technical complexity in all these systems.

Some GB success conditions represented by success factors of other complex systems are also the conditions differentiating GBs from non-GBs. The GB success conditions which are also the differentiating conditions are related to communication, preferences in project team selection, end-users' involvement in defining aspirations, team collaboration, commissioning and fine-tuning, sustainability brief, and project monitoring and control. This implies that a number of success conditions ($n = 8$) represented by success factors of other complex systems are in fact the conditions setting GBs apart from traditional non-GB projects. This also implies that the complexity of GB projects tends to set them apart from non-GB projects.

Table 4.2 Comparison of Green Building success conditions with success factors of other complex projects

Success factors identified in reviewed studies	Complex projects							Sum	Success conditions of Green Building projects
	Global information systems	Large-scale complex projects	Large construction projects	Integrated pest management projects	Complex multinational projects	International development capacity-building projects	Large projects in aerospace and defence sectors		
Commitment of stakeholders			X		X	X		3	Stakeholders' approval of project
Communication (cross-functional communication and clear communication)	X		X				X	3	Communication among project team[a]
Understanding and approval by top management	X		X				X	3	Client's motivation to achieve sustainable outcomes PM team motivation to achieve sustainable outcomes
Adequate funding throughout the project			X				X	2	Accessibility of project funding
Competent project managers			X				X	2	Proficiency of project management (PM) team
Involving cross-functional teams	X		X					2	Preferences in project team selection[a]
Project monitoring and control						X	X	2	Monitoring and controlling the operational performance of building[a] Monitoring of project development Control of project design by the design and sustainability consultant
Motivation of project team							X	1	Project team motivation to achieve sustainable outcomes
Adequacy of project team							X	1	Proficiency of project team
Alignment of personal and organisational interests with project outcomes						X		1	Alignment of team interest with project interest
Availability of resources			X					1	Access to sustainable building materials

(Continued)

Table 4.2 Comparison of Green Building success conditions with success factors of other complex projects (Continued)

Success factors identified in reviewed studies	Complex projects							Sum	Success conditions of Green Building projects
	Global information systems	Large-scale complex projects	Large construction projects	Integrated pest management	Complex multinational projects	International development capacity-building projects	Large projects in aerospace and defence sectors		
Awarding bids to the right designer/contractor			X					1	Preferences in project team selection[a]
Bringing end-users on board early during the project	X							1	End-users' involvement in defining project aspirations[a]
Budget management							X	1	Adequate budget allocation for project development
Clear objectives and scope			X					1	Clarity in communication of project goals
Community involvement			X					1	Stakeholders' approval of project
Comprehensive contract documentation			X					1	Provision of sustainability specifications and other related information in tender
Leadership attributes		X						1	Client's leadership in project Leadership qualities among project team members
Planning							X	1	Rigour of project planning
Practical and continuous training of team				X				1	Educating project team about GB development
Preparation of client's team							X	1	Educating client about sustainability in project Educating end-users and facility management (FM) team about building operation
Project definition							X	1	Setting of a detailed sustainability charter or brief[a]

(Continued)

Table 4.2 Comparison of Green Building success conditions with success factors of other complex projects (Continued)

Success factors identified in reviewed studies	Complex projects							Sum	Success conditions of Green Building projects
	Global information systems	Large-scale complex projects	Large construction projects	Integrated pest management projects	Complex multinational projects	International development capacity-building projects	Large projects in aerospace and defence sectors		
Promoting ownership				X				1	Delegating clear responsibilities to project team
Proper emphasis on the past experience			X					1	Proficiency of project team Proficiency of project client
Decision-making abilities of clients				X				1	Client's rational decision-making
Risk management							X	1	Rigour of risk management
Team collaboration						X		1	Project team collaboration[a]
Verification and testing							X	1	Inspection of project upon construction Execution of commissioning and fine-tuning[a]

Note
a The success condition is also a differentiating condition.

4.3.2 Interpretation of success conditions based on the Transformation-Flow-Value-generation theory

Koskela's theory of production (that is, the TFV theory) sees construction as production from three viewpoints: as a chain of transformations, as the flow of work, and as a generation of value for the customer (Koskela, 2000). To provide a holistic understanding of the application of the TFV theory on GB project success in this section, the findings are interpreted at the level of broad *themes* instead of individual conditions.

In the Transformation view of the TFV theory, production is a conversion of inputs into outputs (Koskela, 2000). According to this view, GB project development is a transformation of labour, information, equipment, and materials into a sustainable building. Production management in this viewpoint equates to decomposing the total transformation into elementary transformations and tasks, acquiring the inputs to these tasks with minimal costs and carrying out the tasks as efficiently as possible (Koskela, 2000). Compared to the Transformation view, the Flow view provides a better understanding of how to avoid the waste of resources. In the Flow view, the basic thrust is to eliminate waste from the flow processes, hence promoting such principles as lead-time reduction, variability reduction, and simplification (Koskela, Howell, Ballard, & Tommelein, 2002). In this view, the GB project development is a flow of resources (such as people, material, and information). A key principle of the Flow view is to reduce the share of non-value-adding activities (that is, waste) in the project (Koskela, 2000). In comparison, the Value-generation view is focused on meeting customer requirements. The view of production as Value-generation has the basic goal of reducing the gap between achieved and best possible value, or in other words achieving the best possible value for the customer (Koskela et al., 2002). In this view, production management is to translate customer needs accurately into a design solution, and then producing products which conform to the specified design (Koskela, 2000). In the Value-generation view, the GB project development is a process of producing a sustainable building which meets the client's requirements. Overall, the TFV theory of production accounts for chains of transformations, flow of work, and Value-generation for the customer.

The TFV framework helps conceptualise the waste in the process of project development and the waste related to the end-product (such as building project). Bølviken, Rooke, and Koskela (2014) provided a taxonomy related to the waste of production in construction according to which the wastes in the Transformation view are material waste, non-optimal use of materials, machinery, energy, or labour; wastes in the Flow view are unnecessary movement of people, unnecessary and inefficient work, waiting, space not used, materials not processed, and unnecessary transportation; and wastes in the Value-generation view are lack of quality and intended use, harmful emissions, and injuries and work-related sickness. The association of success conditions with the TFV theory is interpreted based on the principles of the Transformation, Flow, and Value-generation views and the concepts of waste embedded in these views. The associations of the *themes* of success conditions with the TFV theory of production are shown in Figure 4.1, and the details of these associations are provided in Table 4.3.

Although all the *themes* shown in Figure 4.1 and explained in Table 4.3 are associated with the Value-generation view, some *themes* such as 'Defining project goals' (Theme-7) and 'Design methodology' (Theme-8) are more strongly embedded in this view. This is because these *themes* capture project requirements and convert these requirements into a workable plan (that is, design). Some *themes* which are strongly associated with the Flow view are about the rigour of deliverables (Theme-4), engraving sustainability (Theme-10), information flow (Theme-11), monitoring and control (Theme-12), team collaboration (Theme-16), team commitment (Theme-17), and timeliness of activities (Theme-20). This is because these *themes* strongly

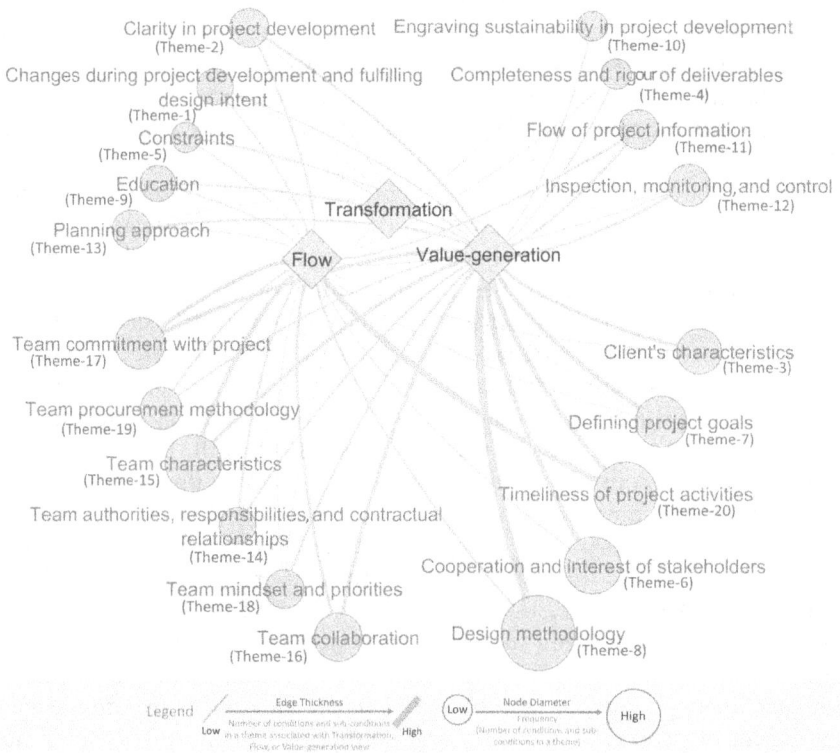

Figure 4.1 Network representing associations of the success condition themes with the Transformation, Flow, and Value-generation views.

Table 4.3 Associations of the success condition themes with the Transformation, Flow, and Value-generation views

Theme number	Theme title	Association of theme with the TFV theory
1	Changes during project development and fulfilling design intent	The fulfilment of design intent helps create value for the client and contributes to the transformation of resources (for example, material and labour) into end-products, in other words, GBs. Moreover, by reduced changes in the project during development, the wasteful activities (for example, project redesign) are reduced, hence associating it with the Flow view
2	Clarity in project development	'Clarity in project development' means that the project team has clearly defined roles, responsibilities, and guidelines to follow. The resulting clarity in a team's vision contributes to the transformation process and reduces the gap between achieved and best possible value. The clarity in vision also reduces the possibility of unnecessary and inefficient work (that is, flow-related waste), hence avoiding the non-value-adding activities
3	Client's characteristics	'Client's characteristics' means that the client as a person or an organisation has the necessary attributes and skill sets to effectively contribute towards successful GB project development. Considering the client's influential role in engaging project team and driving the project development process, the client's skill set can reduce the likelihood of non-value-adding activities. Client's characteristics also affect the project decisions (that is, defining project targets), hence controlling the gap between achieved and best possible value
4	Completeness and rigour of deliverables	'Completeness and rigour of deliverables' is about the level of completion and rigour of deliverables (for example, brief, design, and certification documentation) related to different project stages. Since the deliverables have a high degree of association with the project development process, their rigour and completeness have far-reaching effects on project activities. For instance, well-developed design documentation ensures better construction. Therefore, well-developed deliverables reduce the possibility of unnecessary and inefficient work (that is, flow-related waste) and contribute towards the aspired value for the project client
5	Constraints	'Constraints' are the conditions over which project teams have limited control such as material availability, project funding, and access to the project location. As a result of constraints, the project inputs such as funding and building materials may not be available, and therefore transformation into output (that is, GB project) may be restrained. Constraints can also lead to non-value-adding activities (for example, project redesign) and can also increase the gap between achieved and best possible value

(Continued)

Table 4.3 Associations of the success condition themes with the Transformation, Flow, and Value-generation views *(Continued)*

Theme number	Theme title	Association of theme with the TFV theory
6	Cooperation and interest of stakeholders	'Cooperation and interest of stakeholders' is about the involvement of client, end-users, building control authorities, and neighbourhood community in GB development and operation. Cooperation of these stakeholders helps achieve the aspired project value by easing the pursuit of project goals. Cooperation of some stakeholders (such as the project client, building control authorities, and neighbourhood community) also contributes towards a streamlined project development process, hence the likelihood of non-value-adding activities is reduced
7	Defining project goals	'Defining project goals' is about the project aspirations defined by the project team, FM team, and end-users. Since the achieved value in a project is intrinsically dependent on the aspired value, the better approaches of defining GB aspirations strongly contribute towards value for the project client. Rigorously defined project goals result in realistic aspirations, therefore reducing the occurrence of non-value-adding activities (for example, re-planning and redesigning)
8	Design methodology	'Design methodology' is about the rigorous development of project design to meet the project criteria of sustainability, maintainability, adaptability, reliability, buildability, and innovation. Since design development is a critical activity to transform aspirations into reality, it plays an important role in creating value for the project client. The rigour of design development also reduces the likelihood of errors in design documentation and construction, hence reducing the non-value-adding activities (for example, redesign and reconstruction)
9	Education	'Education' is about raising the GB development and operation-related awareness of the project client, end-users, and project team. The increased understanding of GB resulting from education can align the efforts of the project team and key stakeholders towards successful project development and operation, hence creating value in the project. Since the project team and client are actively involved in the project development process, the increase in their awareness also reduces the Flow-related waste (that is, unnecessary and inefficient work)
10	Engraving sustainability in project development	'Engraving sustainability in project development' is about incorporating sustainability principles and practices in different aspects of project development such as site procurement and construction. Since sustainability in GBs is one of the measures of project value, engraving sustainability in project development means generating value for the project. Moreover, as engraving sustainability also results in the efficient achievement of outcomes (that is, at low cost, with less labour, and fewer resources), it positively associates with the Flow view

(Continued)

Table 4.3 Associations of the success condition themes with the Transformation, Flow, and Value-generation views (*Continued*)

Theme number	Theme title	Association of theme with the TFV theory
11	Flow of project information	'Flow of project information' is about team communication and access to information. A smooth flow of information among the project team and across project stages can reduce the various non-value-adding activities (for example, inefficient design and construction) associated with the lack of information. Moreover, smooth information flow also reduces the risk of not meeting project requirements, hence contributing towards value for the client
12	Inspection, monitoring, and control	'Inspection, monitoring, and control' ensures that the project performs according to the aspirations and value for the client is created. Monitoring of the project also helps avoid such mistakes, which in the later stages of development and operation may need corrections (that is, resulting in non-value-adding activities)
13	Planning approach	'Planning approach' is about rigorously planning for time, cost, and risk in the project. Since better planning helps achieve project aspirations, it positively associates with project value. Project planning is about decomposing the project into smaller units (that is, activities) to which cost, time, and resources can be allocated. Because of this reason, planning is strongly associated with the Transformation view which has a key principle that the total transformation can be decomposed into smaller transformations. By paying attention to details, arranging activities in their logical sequence, and allocating adequate time and resources for activities, the likelihood of errors and mistakes is reduced and therefore non-value-adding activities (that is, inefficient work) are also reduced
14	Team authorities, responsibilities, and contractual relationships	'Team authorities, responsibilities, and contractual relationships' is about delegating responsibilities to the project team. The control and responsibilities delegated to the project team can align their interests towards project goals and therefore create value for the client. Moreover, it also helps reduce non-value-adding activities since the project team has knowledgeable professionals who can efficiently develop a project in case they have the authority to control it
15	Team characteristics	'Team characteristics' is about team proficiency and the ability of the project team to work together. Team characteristics strongly associate with the Flow view since the project team can control the project development and a proficient team can considerably reduce the non-value-adding project activities (for example, project redesign because of incompetent design consultant). Proficiency of the project team also reduces the risk of not meeting project requirements, hence the probability of achieving value for the client is increased

(*Continued*)

Table 4.3 Associations of the success condition themes with the Transformation, Flow, and Value-generation views *(Continued)*

Theme number	Theme title	Association of theme with the TFV theory
16	Team collaboration	'Team collaboration' helps the team in pursuing project goals, therefore achieving value for the client. Team collaboration also leads to more integrated and efficient project development, therefore reducing the non-value-adding activities
17	Team commitment to the project	'Team commitment to the project' is about the commitment of development and operation team to achieve project outcomes. Owing to the important role of the project team in the development process, team commitment eases the pursuit of project goals, therefore resulting in value for the client. Team commitment also results in the rigour of project development (that is, meeting quality standards), and this reduces the non-value-adding activities (that is, rework because of poor quality)
18	Team mindset and priorities	'Team mindset and priorities' is about the viewpoints used by the project team while working together and in achieving project outcomes. Owing to the critical role of the project team in the development process, its priorities significantly determine the achievement of project outcomes, therefore affecting the value generated in the project. The open-mindedness and innovative mindset of the project team result in efficient development approaches, therefore reducing the non-value-adding activities
19	Team procurement methodology	'Team procurement methodology' is about selecting suitable professionals for project activities. Team procurement significantly contributes towards value for the client since the selection of suitable professionals means that the risk of not meeting project requirements is reduced. Moreover, the selection of suitable professionals also means increased opportunities for efficiently developing the project, therefore reducing the non-value-adding activities
20	Timeliness of project activities	'Timeliness of project activities' is about early team engagement, early introduction of project targets, and timeliness of GB certification documentation. The timeliness of activities increases the possibility of meeting project requirements, therefore easing the generation of the project value. In case project activities are timely performed, opportunities for a more efficient development (for example, reduced labour, cost, and material resources) are created, therefore reducing the non-value-adding activities

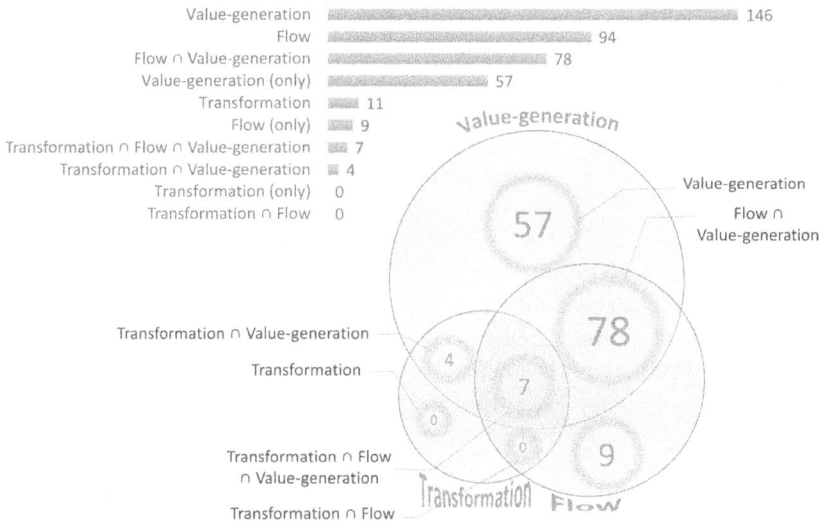

Value-generation   146
Flow   94
Flow ∩ Value-generation   78
Value-generation (only)   57
Transformation   11
Flow (only)   9
Transformation ∩ Flow ∩ Value-generation   7
Transformation ∩ Value-generation   4
Transformation (only)   0
Transformation ∩ Flow   0

Value-generation

Value-generation

Flow ∩
Value-generation

57

78

Transformation ∩ Value-generation

Transformation

4

7

Transformation ∩ Flow
∩ Value-generation

0

0

9

Transformation ∩ Flow

Transformation   Flow

Figure 4.2 Association of success conditions with the Transformation, Flow, and Value-generation views of the TFV theory.

influence project development in terms of the flow of materials, resources, and labour. Unlike Flow and Value-generation, the Transformation view is associated with a low number of *themes*. The *theme* strongly associated with the Transformation view is planning approach (Theme-13), since it is embedded in the concept of breaking the overall transformation (that is, project) into a number of component transformations (that is, activities) which are individually optimised for cost, time, and resources.

For theoretical interpretation of the findings, an overview of the associations of success conditions with the TFV views is provided (Figure 4.2). Following are some key highlights of the associations observed:

- About 94% (146) of success conditions and sub-conditions are associated with the Value-generation view, which means that they directly affect the value for the project client. This is unsurprising since success conditions are about achieving project outcomes, and by doing so value for the project is created.
- About 61% (94) of success conditions and sub-conditions are directly associated with the Flow view. This means that most success conditions are about creating efficiency in work and reducing non-value-adding activities.
- High overlap of 78 success conditions associated with both the Flow and Value-generation indicates that half of the overall success conditions simultaneously add value in the project and contribute towards efficient project development.

- About 7% (11) of success conditions and sub-conditions are associated with the Transformation view. Even these few conditions are not explicitly associated with the Transformation view as they are also associated with Flow and Value-generation views. This shows that even though transformation is an important underlying concept, it does not necessarily have a direct association with many success conditions.

The high association of success conditions with the Value-generation and Flow views means that these two views in the TFV theory are most relevant in explaining the success conditions. This, therefore, indicates that the conditions which critically determine project success are mainly those which add value in a project for the client or reduce the non-value-adding activities.

Although some success conditions are not directly associated with the Transformation, Flow, and Value-generation views, they may, however, have indirect associations with these views. This is because the success conditions are associated with each other in a complex network. Hence, even if condition 'A' does not directly associate with project value, it interrelates with condition 'B', which is directly associated with project value.

Even though success conditions seem to have scarce associations with the Transformation view in the TFV theory, it does not mean that the Transformation view is unimportant for project success. It is possible that aspects related to transformation are so embedded in the project development process that industry professionals do not consider these as success conditions anymore. As Koskela (2000) highlighted, over the years, the role of the transformation model as the foundational theory is forgotten. In practice as well as in research settings, the transformation concept is most often implicit, and when made explicit, it is rarely treated as a testable and discussible theory. This indicates that although the transformation is an important underlying view of production in the TFV theory, it has limited use in interpreting success conditions of GB projects.

4.3.2.1 Relevance and limitations of using the TFV theory for research on Green Building project success

The high degree of association of identified success conditions with the Flow and Value-generation views shows that the TFV theory is a relevant theoretical lens to interpret GB project success. This theory may also be used for interpreting the success of other construction projects.

Overall, in this study, the concept of GB project success is found to be much more associated with the Value-generation and Flow views than the Transformation view. As highlighted by research findings, success in GB projects creates value for not only the environment and the client but also for the construction industry and society. The definition of 'customer' in the Value-generation view of the TFV theory only includes the project clients, occupants, and to a limited extent the environment. To accurately assess value and waste, the concept of value must be extended to natural resources (living systems) and human resources (social and cultural systems) that are the basis of human existence (Hawken, Lovins, & Lovins, 2013). If all resource types are not considered, delivery factors that may affect the environmental or 'green' values to the customer (for example, environmental burdens in operation, service life, risk of deterioration, convertibility, and flexibility) are overlooked (Klotz, Horman, & Bodenschatz, 2007). For GB projects, it is particularly important to consider society and the environment as additional customers. Johansson and Sundin (2014) also emphasised the need for broadening the definition of the 'customer' to include society in general. Future research needs to rearticulate the definition of 'customer' in the Value-generation view to represent the variety of stakeholders benefitting from sustainable project development. As society is intrinsically part of a global system, Value-generation must be considered in relation to the external environment and social problems (Salvatierra-Garrido, Pasquire, & Thorpe, 2010). Understanding and making this collective value tangible in the briefing and design phases can be pivotal in delivering value and defining waste (Novak, 2012).

Although the TFV theory has a focus on reducing waste, the elimination of waste may not always be desirable when it comes to complex projects such as GBs. Waste is mostly defined in terms of short-term, operational issues. In comparison, projects may involve higher level purposes which may be more important than waste minimisation, such as learning, maintaining the production system in working order, and avoiding catastrophic consequences for the wider world (Bertelsen & Koskela, 2005). For instance, GB projects involve innovation, and while this results in learning opportunities and the likelihood of improved project development practices, it may also result in the waste of resources. As Smart et al. (2003) have argued, waste minimisation cannot be the only or not even the most important pursuit in all situations. The apparent wasteful features of a complex system such as a GB project can be important for the survival and existence of the system and can be associated with long-term benefits. Bak (1997) specifically studied the errors, faults, and accidents as sources for evolution in the form of learning in complex systems. Faults in a system result in a learning process, and this reduces the likelihood of big problems. For instance, a significant crisis in Toyota's Aisin brake valve plant resulting from fire damage was controlled because of the emergent system properties of Toyota's supply chain (Bertelsen & Koskela, 2005). Considering GBs as complex projects, the waste concept originating from the TFV theory may not provide a complete interpretation for success conditions.

While the association of the TFV views with successful GB project development has direct theoretical implications, it also has practical implications. For instance, looking at GB project development from the viewpoint of TFV, project managers would be aware that their decisions on a certain project aspect are more driven towards the Value-generation as compared to the Flow or Transformation. Knowing this can have much value as it creates the opportunity of balancing the project development between Transformation, Flow, and Value-generation. As Koskela (2000) also emphasised, it is not enough to carry out task, flow, and value management as separate functions. Instead, they must be balanced, and their interactions must be controlled to avoid anomalies. The TFV theory predicts that a production system where the operational principles of all three domains are implemented at all levels of managerial action (design, control, and improvement) should have a better performance than one where principles are implemented less comprehensively.

4.4 Comparing success conditions with credits of Green Building certification systems

Credits of GB certification systems are the criteria against which the sustainability of GB projects is assessed, and a certificate is awarded. In the case of GBs, project teams and clients are highly driven towards sustainability certification. In case a success condition occurs as a credit in a certification system, it implies a recognition of its importance in the practice of GB development. Considering the importance of credits, the success conditions and sub-conditions identified in this study are compared with the credits of four GB certification systems, namely Leadership in Energy and Environmental Design (LEED) system by the US Green Building Council, Building Research Establishment Environmental Assessment Method (BREEAM) by the UK Building Research Establishment, Green Star system by Green Building Council of Australia, and Living Building Challenge system by International Living Future Institute. The Australia-based Green Star system is used since the majority of GB professionals who participated in this study through semi-structured interviews (n = 32, 42%) are based in Australia. LEED system originating in the USA in 1998 and BREEAM established in 1990 in the UK are used for analysis as these are the two most popular certification systems worldwide. Another reason behind the reason for using BREEAM is that a significant number of interview participants (n = 13, 17%) are based in the UK. While the popular certification systems such as LEED, BREEAM, and Green Star have a focus towards reducing environmental footprint and improving the indoor environment, they lack emphasis towards a regenerative built environment. To complement this aspect, the Living Building Challenge system is used in analysis since it has a goal of creating a regenerative built environment (Thomas, 2016).

The certification systems are comprised of multiple frameworks focused on different project types and life cycle stages. To compare with the findings in this study, the LEED frameworks used are 'Building Design and Construction' and 'Building Operations and Maintenance'; the BREEAM framework used is 'UK New Construction'; the Living Building Challenge framework used is 'Buildings'; and the Green Star frameworks used are 'Design & As Built' and 'Performance'. The associations of success conditions with credits of certification systems are shown in Figure 4.3.

Among the overall 155 success conditions and sub-conditions identified in this study, 16 conditions and 23 sub-conditions are associated with 11 credits of Green Star, two credits of Living Building Challenge, 24 credits of LEED, and 20 credits of BREEAM. Hence, there is the highest association of identified conditions with the LEED and BREEAM GB accreditations. By association, it means that the credits of certification systems provide a complete or partial representation of particular success conditions. For instance, 'Use of innovative design approach' is a success condition associated with 'innovation' credit of the LEED system. Some success conditions highly associated with the certification systems are 'Use of energy-oriented design approach', 'Monitoring and controlling operational performance of building', 'Execution of commissioning and fine-tuning', 'Procurement of project site based on sustainability goals', and 'Life-cycle based project development approach'. The credits in the four certification systems do not represent 62 (79%) conditions and 70 (75%) sub-conditions. This lack of representation indicates that the certification systems need to account for more number of success conditions.

4.4.1 Role of certification systems in the implementation of success conditions

GB projects can perform better in case the project management aspects (such as good team communication) occur in GB certification systems (He, Kvan, Liu, & Li, 2018). The success conditions of GB projects as identified in this study can occur as credits in GB certification systems promoting effective GB development practices. Typically, the prime focus of GB certification systems is to evaluate the sustainability performance of a GB project. Therefore, the certification systems are more related to success criteria than success conditions. The representation of success conditions by credits of certification systems, although scarce, suggests that certification systems do acknowledge the importance of success conditions to some extent. Based on the purpose and operational nature of certification systems, it can be argued that the critical success conditions should be included in certification systems as credits. World over, the trend of certifying GB development prevails (Petrullo, Jones, Morton, & Lorenz, 2018). Hence, in case the certification systems encapsulate success conditions, the likelihood of successful GB development will increase.

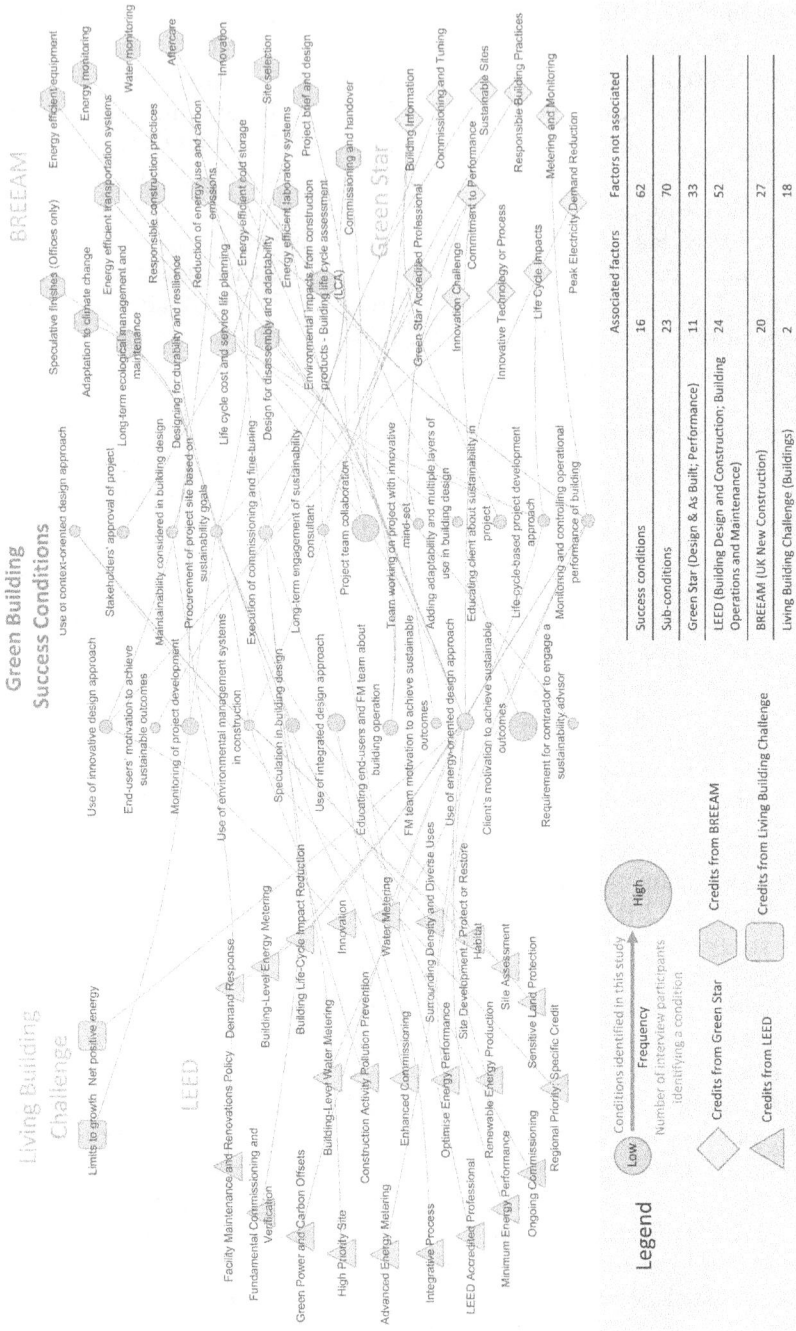

Figure 4.3 Association of success conditions with credits of Green Building certification systems.

Table 4.4 Associations of identified success conditions with credits occurring in Green Building certification systems

Theme	Sr. no. and sub-conditions	Identified success conditions and sub-conditions	Green Star Design and As Built	Performance Credit	Criteria (LBC)	LEED Design + construction	Operations and maintenance	Criteria
Cooperation and interest of stakeholders	17	Client's motivation to achieve sustainable outcomes	X	X — Commitment to performance				
	19.1	End-users' motivation to achieve sustainable outcomes	X	X — Commitment to performance				
Design methodology	26.05	Life-cycle-based project development approach	X	Life cycle impacts		X		Building life cycle impact reduction
	26.06	Maintainability considered in building design					X	Facility maintenance and renovations policy
	26.11	Use of context-oriented design approach				X	X	Site development – protect or restore habitat
						X	X	Regional priority: specific credit
						X		Site assessment
	26.12	Use of energy-oriented design approach	X	X — Peak electricity demand reduction	Net positive energy	X	X	Demand response
						X	X	Green power and carbon offsets
						X	X	Minimum energy performance
						X	X	Optimise energy performance
						X		Renewable energy production
	26.14	Use of innovative design approach	X	Innovative technology or process		X	X	Innovation
	26.15	Use of integrated design approach				X		Integrative process
Education	28	Educating client about sustainability in project	X	X — Building information				
	29	Educating end-users and FM team about building operation	X	X — Building information				

(Continued)

Table 4.4 Associations of identified success conditions with credits occurring in Green Building certification systems (Continued)

Theme	Sr. no.	Identified success conditions and sub-conditions	Green Star — Design and As Built	Green Star — Performance	Green Star — Credit	Criteria (LBC)	LEED — Design + construction	LEED — Operations and maintenance	LEED — Criteria
Engraving sustainability in project development	32	Procurement of project site based on sustainability goals	X		Sustainable sites	Limits to growth	X		High priority site
							X		Sensitive land protection
							X		Surrounding density and diverse uses
	33	Use of environmental management systems in construction	X		Responsible building practices		X		Construction activity pollution prevention
Inspection, monitoring, and control	38.1	Execution of commissioning and fine-tuning	X	X	Commissioning and tuning		X		Enhanced commissioning
							X		Fundamental commissioning and verification
								X	Ongoing commissioning
	39	Monitoring and controlling operational performance of building	X	X	Metering and monitoring			X	Building-level water metering
							X	X	Water metering
							X	X	Advanced energy metering
							X	X	Building-level energy metering
Team commitment to the project	59	FM team motivation to achieve sustainable outcomes	X	X	Commitment to performance				
Team mindset and priorities	65	Team working on project with innovative mindset	X	X	Innovation challenge				
Team procurement methodology	68.1	Long-term engagement of sustainability consultant	X	X	Green Star-accredited professional		X	X	LEED-accredited professional

Providing a quantitative measure of the criticality of success conditions requires a data-intensive investigation and can be enabled by GB certification systems. The importance of success conditions has been analysed through their frequency of occurrence in interviews. Similarly, the network attributes of success conditions also provide a measure of their importance. Previous studies on GB project success have used regression analysis on multi-case studies to determine the criticality of success factors. Instead of being a strictly explanatory inquiry of associations among conditions and criteria, this study is an exploratory inquiry of success conditions. Unlike exploratory design, the explanatory approaches can provide a quantitative measure of the associations and therefore can be used to evaluate the criticality of conditions based on their associations with success criteria. Owing to a large number of success conditions, the data collection for such explanatory research can be a highly laborious and resource-intensive exercise. This is where GB certification systems can be helpful not only in implementing best practices (such as success conditions) but also in investigating success conditions.

Although a direct adoption of study findings in GB certification systems is possible, such incorporation can also benefit from additional research to determine the criticality of success conditions identified in this study. For the potential uptake of success conditions by GB certification systems, two scenarios are proposed as shown in Figure 4.4. In Scenario-1, a certification system adopts the success conditions identified in this study and other studies conducted on GB project success. In this scenario, the certification system considers those conditions as critical. Based on the criticality of conditions, they may be assigned different weights in the certification system. While Scenario-1 is simple and readily applicable, it faces the challenge

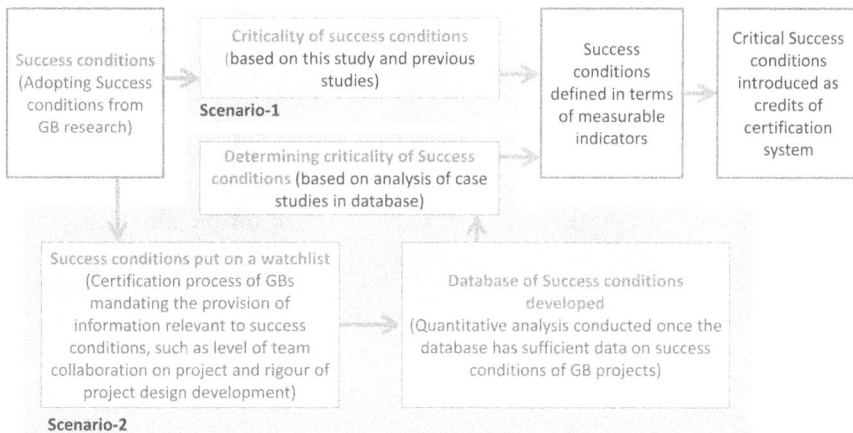

Figure 4.4 Scenarios for the adoption of success conditions by Green Building certification systems.

of determining the criticality of success conditions. This is because the research on GB project success, including this study, does not rigorously account for the relative importance of a large number of success conditions. To complement this limitation, Scenario-2 is proposed where the certification system can introduce additional steps, that is a research pathway. As a first step, a list of success conditions from this study and similar studies on GBs is prepared. The values of these conditions are recorded for new projects accredited with GB certifications, and once enough data from multiple case studies is available for quantitative analysis, the criticality of conditions for project success is determined. The critical conditions can then be considered as credits within the certification system. Hence, in Scenario-2, a GB certification system can help conduct a rigorous inquiry regarding the criticality of success conditions. An important aspect of both these scenarios is that the success conditions need to be converted into measurable parameters (for example, 'collaboration', a success condition can be measured in terms of the number of formal and informal project team meetings). By the application of Scenario-1 and Scenario-2 (shown in Figure 4.4) through GB certifications systems, it is possible to accomplish both the verification and the implementation of success conditions in GB projects.

4.5 Research limitations and suggestions for future studies

A detailed inquiry regarding the successful development of GB projects is conducted in this study. Owing to the limitations in resources and time availability, some limitations exist in terms of the research paradigm, data collection, and the interpretation of findings. These limitations include the limited sample size, limitations in generalisability, time-bound relevance of findings, and comparability with other research on this topic. A brief account of these limitations is provided in this section, as well as the suggestions to address them. The future research areas are also applicable to other types of projects beyond GB projects.

Since the GB market is still in the early stages of its evolution, the number of GB projects and the experience of building professionals regarding GB development are limited. As this study is based on the knowledge and experience of industry professionals regarding GBs, the early stages of GB market evolution limits the research findings. Due to the limited number of GBs, the expert sampling of industry professionals is also limited. To provide specificity, this study has only involved GB professionals having the experience of being involved in green office projects, and this further limits the sample set.

During interviews, some participants have also mentioned success conditions for projects other than office buildings; however, the majority of research findings are related to office buildings. Hence, the overall research findings are related to green office projects and may not be generalised to other building types.

While this study provides a detailed account of the success conditions, it is acknowledged that the study is not exhaustive in terms of the factors and their interrelationships identified. Future studies conducted with the approaches proposed in this section can help identify the success conditions which remained undiscovered in this study, and this can further enrich the definitions and associations identified in this study.

4.5.1 Limitations related to research paradigm and epistemology

Since this study embraces the critical realism paradigm, it adopts a realist ontology and epistemic relativism. Epistemic relativism is based on the understanding that knowledge is articulated from various standpoints according to various influences and interests and is transformed by human activity. While knowledge is context dependent, the representations of the world are historical, perspectival, and fallible (Archer, Decoteau, Gorski, Little, & Porpora, 2016). The study has the stance that reality exists independent of the human conception; however, we need human perception to know about reality. While the perception of GB professionals informing the study reported in this book is justified using the critical realism paradigm, the study may lack comparability with research on similar topics conducted using other paradigms, such as constructivism and interpretivism. Studies conducted on the same topic while using positivist, constructivist, and interpretivist paradigms may result in some findings different from the results reported in this study.

As this study is based on epistemic relativism, it acknowledges that the research findings have a historic context. The study is based on cross-sectional data. Interviews for this study were conducted from September 2017 to February 2019.

Since the GB market is rapidly evolving, the innovations in this sector may limit the applicability of some research findings in future. While acknowledging time as an important function for the research findings, the study presents the findings such that they are comparable with similar studies in the past and future. Some of the findings may have limited application in the future when significant changes occur in the context of GB project development. According to epistemic relativism, knowledge is articulated from various standpoints and the representations of the world are not only historical, but also perspectival. This study is basically developed on the knowledge acquired from GB service-providers and therefore employs their perspective of the GB project success. The interview participants had different roles in the development of GB projects, with the two most common roles among the participants being 'design consultant' ($n = 19$) and 'sustainability consultant' ($n = 22$). From a total of 75 interview participants, only two were representative of end-users and project clients. This indicates that findings of this study are almost entirely limited to the knowledge acquired from GB service-providers, such as design consultants and sustainability consultants. Even though project clients, end-users, and regulatory organisations do not have such active contribution in GB project development as in the case of service-providers, a study focussed on project clients and end-users for data collection may uncover some aspects of project success left unidentified in this study.

4.5.2 Interview-based data collection for research on Green Building project success

This study is informed by data collection from semi-structured interviews. Most of the interviews were conducted in-person, and for this reason, the researcher had to travel extensively. Owing to the logistical constraints, some interviews were also conducted over the telephone. Face-to-face interviewees were engaged better with the researcher and provided more in-depth discussion related to interview themes compared to those in telephone interviews. Owing to the effort-intensive data collection process, the sample size for this study was limited to 75 interview participants. Future studies can consider the regions not covered in this inquiry because of its limited scope. Future studies also need to investigate success conditions for GBs belonging to residential, industrial, retail, educational, hospitality, and health sector.

For research on GB projects, the associations between conditions and criteria are also important along with the identification of success conditions and criteria. The identification of conditions and criteria, as well as their associations, depends on the number of interview participants in case a research inquiry is exploratory and employs a fieldwork approach, as used in this study. With the higher number of participants, a greater number of conditions are identified, and a more in-depth understanding of associations is provided. The identification of these conditions is also subject to the context in which interview participants are based (such as the region of belonging or professional role) as explained in greater detail in Annex B. Hence, for such a research inquiry, it is suitable to have not only a larger number of participants but also more participants from diverse contexts. Although this study has engaged 75 interview participants belonging to six countries, there are still limitations in data collection owing to the limited research resources. For future research on GB project success conducted with the same or similar aim, it is suggested to complement the limitations of this study by

- Interviewing a greater number of project clients, facility management professionals, and end-users
- Interviewing a greater number of third-party stakeholders, such as local government, GB certification authorities, building inspectors, and neighbourhood community

- Interviewing GB professionals in developing economies (for example, India, Pakistan, and Nigeria) and developed economies (for example, the USA and Germany)
- Interviewing GB professionals from regions where the GB development faces special climatic constraints, for example, heat, water scarcity, or cultural constraints

4.5.3 The sequence of general and specific themes in semi-structured interviews

The in-depth interviews in this study had a general *theme* in which the participants were asked to reflect on the conditions which make GBs successful. Interviews also had a specific theme, in which the participants discussed the conditions which contributed to the performance of successful and failed projects in their portfolios. Both the general and specific themes had their usefulness in data collection, and they enabled the interviewer to actively engage the participants. For instance, in case a participant was unable to differentiate between success conditions and success criteria, it was a useful approach to first discuss the specific theme followed by the general theme. Upon recalling the details of a project, a participant could talk about conditions and criteria with more clarity. While this approach of 'Specific to General' had its advantages, its limitation is that a participant may end up giving the same answer for a general *theme* as s/he previously provided for a specific theme. This may occur as the specific themes can trigger an availability heuristic which is a mental shortcut relying on immediate examples that a person recalls when evaluating a specific topic, concept, method, or decision. The availability heuristic operates on the notion that if something can be recalled, it must be important to them or at least be more important than alternative solutions, which are not as readily recalled (Folkes, 1988).

The 'General to Specific' approach can help avoid the availability heuristic while the 'Specific to General' approach can result in more clarity of concepts for an interview participant. While understanding the potential benefits and pitfalls of these approaches, the author/researcher in this study referred to both these approaches under different interview circumstances for effective data collection. Hence, both the approaches had their usefulness and limitations, and their appropriate use was subject to the researcher's judgment, time available for conducting an interview, and the cognitive abilities of participants.

4.5.4 Green Building project success research: Challenges of data collection

To develop the knowledge of GB project development based on the perceptions of industry professionals, future studies will continue to use online surveys and interviews; however, they may need to cater for the following challenges:

- For two factors, two associations are to be identified, and for six factors, these associations rise to 30, indicating an exponential increase in the number of associations with the increase in the number of inquired factors. This highlights the impracticality of investigating associations for a large number of factors in online surveys or interviews.
- The list of success conditions created in this study is based on the collective knowledge of 75 interview participants belonging to six countries. This highlights that a person participating in such a study through survey or interviews is unlikely to possess the knowledge and experience to reflect on the associations among all the factors being inquired.

To some extent, the second challenge can be addressed using the modularisation approach to knowledge acquisition, as also used in this study. While using this approach, the author/researcher actively adapted the interview questions to the experience and professional role of interview participants. This way, the participants answered for only those aspects of GB development regarding which they had most knowledge and experience. Details regarding this approach are provided in Annex A (Section 6.3).

4.5.5 Longitudinal study of Green Building project success

Based on the philosophical paradigm of critical realism, it is believed that the research findings are time dependent and therefore have a historic context. The findings of this study may change with time, and in this regard, the thoughts of an interview participant *(AU-M-14)* are worth mentioning,

> Initially, we used to think of success in the way that how we can change the design industry and bring innovation into the design of Green Buildings. This was around ten years ago. These days, it is a given …. Now, the expectation is that, how we can operate the buildings efficiently and there needs some work to be done in this regard.

This observation implies that the perceptions of GB success conditions are subject to time, and this is an additional validation that critical realism is a well-suited paradigm for this study. Recognising time as an important attribute of this research, adopting a longitudinal research approach can yield important findings. One way of enabling a longitudinal study is by interviewing the same GB professionals as previously interviewed or in case they are inaccessible, interviewing the professionals from the same contexts to which the participants of this study belonged. Such an approach can help understand how the perception of project success is affected by time. Such an inquiry of GB project success is missing from the literature, and by conducting it, a significant contribution to the theory of GB project success can be made (Ahmad, Aibinu, & Stephan, 2019; Korkmaz, Horman, & Riley, 2009). Such a longitudinal study of GB project success can also contribute towards the inquiry of GBs in terms of innovation diffusion. By considering GBs as innovation in the construction industry, the longitudinal study of GB project success can inform how the diffusion of innovation impacts the perceptions of project success.

4.5.6 Developing appropriate scope for research on Green Building project success

Since this study reports a large number of factors, the individual factors are reported as succinctly as possible. In this study, many factors have been identified in terms of success conditions (conditions = 73; sub-conditions = 82) and differentiating conditions (28). The factors identified in this study involve a description of their importance and their interrelationships. The explanation regarding the identified factors is an important aspect of this study. Even though each factor merits a detailed analysis and reporting, it is challenging to concisely report a large number of factors. Furthermore, the detailed analysis of individual factors may result in a much broader scope of research as compared to the scope of this study. Under these constraints, the factors in this study are presented as concisely as possible.

For future studies in the area of GB project success, it is suggested to give due consideration to project scope while developing research design. In terms of deciding research scope, there are two extreme approaches – scaled-down and detailed - and a balanced approach as shown in Figure 4.5. The research conducted in this study lies between balanced research approach and detailed research approach.

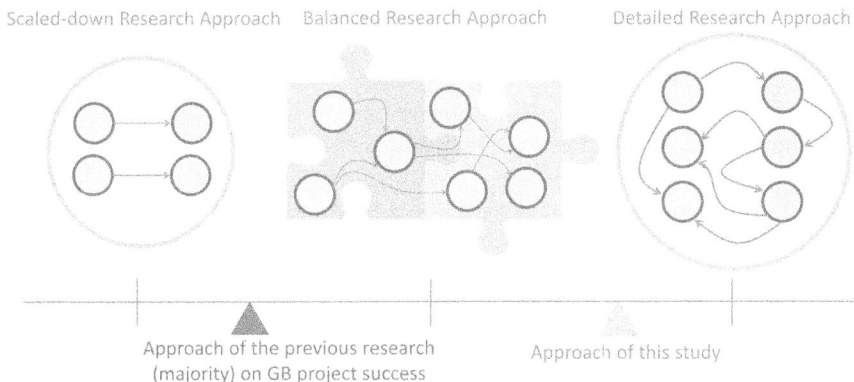

Figure 4.5 Approaches for research on Green Building project success.

On one end, there is an approach of scaling down the research scope in terms of things like building type, stage of the project (for example, development or operation), focused stakeholders (for instance, success factors for the project manager or client), and region. The benefits of scaled-down studies are that the data collection and analysis are reasonably easier to conduct; more rigorous analyses of factors (such as success conditions and criteria) and their associations are possible within limited resources; and the analysis findings are easier to report. On the downside, this approach is reductionist and ignores the complex nature of GB project development and operation. Upon scaling the research scope down to one stage of a project or a particular stakeholder, either many factors are consolidated or their sheer number is reduced, resulting in a blind side towards the factor associations.

At the other end is the approach of developing detailed research on GB project success. Such an approach thrives on the discovery of a greater number of factors and factor associations. Therefore, such research can benefit from a broader scope in terms of project stages, stakeholder opinions, regional diversity, and the like. Since this approach focuses on unfolding complex systems, it is suitable for inquiring GB project success, given that these projects have socio-technical complexity (Rohracher, 2001). Considering complex systems, the more factors and associations are discovered, the more in-depth understanding of a system's working is developed. The broader scope because of the detailed research approach means an opportunity of knowing more about factors and their associations, therefore not only broadening but also deepening the insight of a complex system, like GB project success. A major limitation of this research approach is in terms of its unprecedented scope, requiring an excessively large amount of resources.

Between the two extremes of scaled-down approach and detailed research approach lies a balanced research approach. In such an approach, studies are conducted with a limited scope such that they are consistent with previous studies and are reported in a way that studies on similar patterns can be conducted in the future. To understand factor associations, the balanced approach limits reductionism. Such studies need to be reported so that the subsequent studies can start where they end. In practice, previous research on GB project success is conducted somewhere between scaled-down highly reductionist approach and a balanced approach. Studies conducted on balanced approach can be considered as parts of a jigsaw puzzle: while the individual studies do not account for a large number of factors and associations, in combination with other studies they can provide a holistic as well as an in-depth understanding of GB project success. Studies conducted in the future could adopt a balanced research approach. By limiting these studies to the scope of individual project stages (for example, design, construction, and operation), different building types, regions, and stakeholders, the studies will have a manageable number of factors which can be analysed in detail and reported adequately at the level of individual factors. For future studies to successfully follow the balanced approach, it is necessary to disseminate not only the interpretation of the findings but also the raw data so that a better integration among the findings of these studies can be achieved.

4.6 Summary

In this chapter, the success conditions are interpreted in terms of the theory and practice of GB project development. The effect of the interview themes and attributes of interview participants on the identified success conditions is addressed. It is realised that the identification of success conditions is affected by the themes used in interviews as well as the regional belonging, experience, and role of industry experts in GB projects.

The literature on project success of complex systems is reviewed to compare the findings of GB project success with success factors of complex projects belonging to different industries. A considerable match of success conditions across GB projects and other complex projects is realised, indicating that because of the socio-technical complexity of these projects, there are some common success conditions across these systems.

This chapter has also analysed GB success conditions in terms of the TFV theory. The success factor interpretation using TFV theory has shown that the conditions within all the 20 success themes are found to reduce the non-value-adding activities in project development, create value for the client, and in some cases also result in transformation. Most of the success conditions are associated with Flow and Value-generation views. Very few success conditions are associated with the Transformation view.

To address the practical implications of research findings, a comparison of the success conditions with the credits of GB certification systems (that is, Living Building Challenge, LEED, BREEAM, and Green Star) is provided. From the overall 155 success conditions and sub-conditions identified in this study, 16 conditions and 23 sub-conditions are associated with 11 credits of Green Star, two credits of Living Building Challenge, 24 credits of LEED, and 20 credits of BREEAM. This helps understand how some success conditions are already being implemented by GB certification systems. The potential of using GB certification systems for the adoption of GB success conditions is also explored.

The chapter also addressed the limitations associated with the findings of this study. This study embraces the critical realism paradigm and acknowledges that the studies employing positivist, constructivist, and interpretivist paradigms may have some differences from the findings of this study. Future studies are suggested to employ more data collection from service-seekers (that is, project clients, FM managers, and end-users) and consider developing economies and regions with climatic challenges. The need for longitudinal research on the topic of GB project success is suggested. Approaches for developing appropriate research scope for future studies are also discussed.

The next chapter concludes the findings of this study by providing a detailed account of the stakeholders affected by GB success conditions and provides detailed suggestions for key stakeholders of GBs to ensure the likelihood of project success.

References

Ahmad, T., Aibinu, A. A., & Stephan, A. (2019). Managing green building development – A review of current state of research and future directions. *Building and Environment, 155,* 83–104. doi:10.1016/j.buildenv.2019.03.034

Archer, M., Decoteau, C., Gorski, P., Little, D., & Porpora, D. (2016). *What is critical realism?* Retrieved from http://www.asatheory.org/current-newsletter-online/what-is-critical-realism

Bak, P. (1997). How nature works: The science of self-organized criticality. *American Journal of Physics, 65,* 579. doi:10.1119/1.18610

Bertelsen, S., & Koskela, L. (2005). *Approaches to managing complexity in project production.* Paper presented at the 13th International Group for Lean Construction Conference (IGLC), Sydney, Australia.

Biehl, M. (2007). Success factors for implementing global information systems. *Communications of the ACM, 50*(1), 52–58. doi:10.1145/1188913.1188917

Bølviken, T., Rooke, J., & Koskela, L. (2014, June 25–27). *The wastes of production in construction – A TFV based taxonomy.* Paper presented at the 22nd Annual Conference of the International Group for Lean Construction, Oslo, Norway.

Duy Nguyen, L., Ogunlana, S. O., & Thi Xuan Lan, D. (2004). A study on project success factors in large construction projects in Vietnam. *Engineering, Construction and Architectural Management, 11*(6), 404–413. doi:10.1108/09699980410570166

Folkes, V. S. (1988). The availability heuristic and perceived risk. *Journal of Consumer Research, 15*(1), 13–23. doi:10.1086/209141

Hawken, P., Lovins, A. B., & Lovins, L. H. (2013). *Natural capitalism: The next industrial revolution.* London: Routledge.

He, Y., Kvan, T., Liu, M., & Li, B. (2018). How green building rating systems affect designing green. *Building and Environment, 133,* 19–31. doi: 10.1016/j.buildenv.2018.02.007

Ika, L. A., & Donnelly, J. (2017). Success conditions for international development capacity building projects. *International Journal of Project Management, 35*(1), 44–63. doi:10.1016/j.ijproman.2016.10.005

Johansson, G., & Sundin, E. (2014). Lean and green product development: Two sides of the same coin? *Journal of Cleaner Production, 85,* 104–121. doi:10.1016/j.jclepro.2014.04.005

Klotz, L. E., Horman, M., & Bodenschatz, M. (2007). A lean modeling protocol for evaluating green project delivery. *Lean Construction Journal, 3*(1), 1–18. Retrieved from https://www.leanconstruction.org/learning/publications/lean-construction-journal/

Korkmaz, S., Horman, M., & Riley, D. (2009). *Key attributes of a longitudinal study of green project delivery.* Paper presented at the Construction Research Congress, ASCE, Seattle, WA.

Koskela, L. (2000). *An exploration towards a production theory and its application to construction.* (PhD dissertation). Helsinki University of Technology.

Koskela, L., Howell, G., Ballard, G., & Tommelein, I. (2002). The foundations of lean construction. *Design and Construction: Building in Value, 291,* 211–226.

Novak, V. M. (2012). *Value paradigm: Revealing synergy between lean and sustainability.* Paper presented at the 20th Conference of the International Group for Lean Construction, San Diego, CA.

Petrullo, M., Jones, S., Morton, B., & Lorenz, A. (2018). World green building trends 2018—Smart market report. *Dodge Data & Analytics.* Retrieved from https://www.worldgbc.org/sites/default/files/World%20Green%20Building%20 Trends%202018%20SMR%20FINAL%2010-11.pdf

Pisarski, A., Chang, A., Ashkanasy, N., Zolin, R., Mazur, A. K., Jordan, P. … Hatcher, C. A. (2011). The contribution of leadership attributes to large scale, complex project success. Paper presented at the 2011 Academy of Management Annual Meeting, San Antonio, TX.

Rodriguez-Segura, E., Ortiz-Marcos, I., Romero, J. J., & Tafur-Segura, J. (2016). *Critical success factors in large projects in the aerospace and defense sectors* (0148-2963). Retrieved from https://dspace.mit.edu/bitstream/handle/1721.1/2010/ SWP-1297-08770929-CISR-085.pdf?sequence=1

Rohracher, H. (2001). Managing the technological transition to sustainable construction of buildings: A socio-technical perspective. *Technology Analysis & Strategic Management, 13*(1), 137–150. doi:10.1080/09537320120040491

Salvatierra-Garrido, J., Pasquire, C., & Thorpe, T. (2010). *Critical review of the concept of value in lean construction theory.* Paper presented at the 18th Annual Conference of the International Group for Lean Construction, Haifa, Israel.

Schmidt, P., Stiefel, J., & Hürlimann, M. (1997). *Extension of complex issues: Success factors in integrated pest management.* St. Gallen, Switzerland: Skat.

Smart, P., Tranfield, D., Deasley, P., Levene, R., Rowe, A., & Corley, J. (2003). Integrating 'lean' and 'high reliability' thinking. *Proceedings of the Institution of Mechanical Engineers, Part B: Journal of Engineering Manufacture, 217*(5), 733–739. doi:10.1243/095440503322011489

Thamhain, H. J. (2013). Commitment as a critical success factor for managing complex multinational projects. *International Journal of Innovation and Technology Management, 10*(04), 1350016. doi:10.1142/S0219877013500168

Thomas, M. A. (2016). *The living building challenge: Roots and rise of the World's greenest standard.* Fort Worth, TX: Ecotone Publishing.

Chapter 5

Green Building success and project stakeholders

5.1 Introduction

Who should read this chapter

Read this chapter, if you are interested in the following:

- Understanding the stakeholders responsible for and affected by success conditions
- Suggestions related to GB stakeholders to ensure successful GB development

Project success being a function of social attributes is strongly driven by the actions of project stakeholders. This chapter is about exploring the role of project stakeholders regarding GB success conditions. First, a detailed account of the stakeholder responsible for and affected by GB success conditions is provided. This is followed by practical implications regarding the use of success conditions by stakeholders. The potential ownership of success conditions by project stakeholders is discussed. Lastly, a detailed account of the suggestions for GB project stakeholders is provided to ensure effective application of success conditions in GB projects.

5.2 Project stakeholders responsible for and affected by success conditions

A word frequency analysis was conducted on interview data regarding success conditions, and the most used words (as shown in Figure 5.1) include green, building, project, sustainability, team, client, consultant, contractor, development, energy, performance, people, requirements, management, stage, certification, targets, construction, cost, developer, and successful. It shows that a sizable number of participants refer to project stakeholders while discussing GB success conditions. It is important for theoretical and

DOI: 10.1201/9781003322740-7

Figure 5.1 Word cloud summarising the responses to the interview theme of success conditions.

practical reasons to know the stakeholders responsible for and affected by success conditions. In practice, by knowing the important stakeholders for a success condition, that condition can be assigned to stakeholders who control the process of project development. In theory, by knowing the stakeholders affecting and affected by a condition, it is possible to determine the level of significance and complexity associated with a success condition.

In the semi-structured interviews, the term 'client' refers to the party for whom the project procurement takes place, and for whom services of the project team are provided. A project client can be the owner of a GB project, a developer, an investor, and in some cases also the end-user. While a client can be an end-user, the facility management (FM) team can also be the end-user. While a considerable number of success conditions are associated with the overall project team, many conditions are also associated with the specialist groups of the overall team such as the sustainability team and design team. Within these specialist teams, 'Contractor team' includes both the head contractor team and the sub-contractors, and the 'Design team' includes both the architectural design consultants and the MEP design consultants. The term 'External Stakeholders' refers to GB certification organisations (for example USGBC and GBCA), neighbourhood community, and building control authorities.

The identified success conditions are analysed to identify the project stakeholders they are related to. The associations of conditions with project stakeholders are shown in Table 5.1, while the associations of the broader *themes* of success condition with project stakeholders are shown in Figure 5.2. Alongside suppliers and external stakeholders, there are three sets of stakeholders in Figure 5.2. Within the project team set, there is 'Overall Project team' alongside specialist teams such as 'Project Management (PM) team' and 'Commissioning team'. The reason for such distribution is that some success conditions are exclusively related to specialist teams, while some other conditions are related to the overall project team. The highest number of conditions (number of conditions = 43; 28%) is associated with the overall project team. Some conditions are specifically associated with the specialist teams such as design team ($n = 32$; 21%), contractor team ($n = 18$; 12%), sustainability consultant ($n = 13$; 8%), project management (PM) team ($n = 11$; 7%), and commissioning team ($n = 2$; 1%). Following the overall project team, the highest association of conditions is with the project

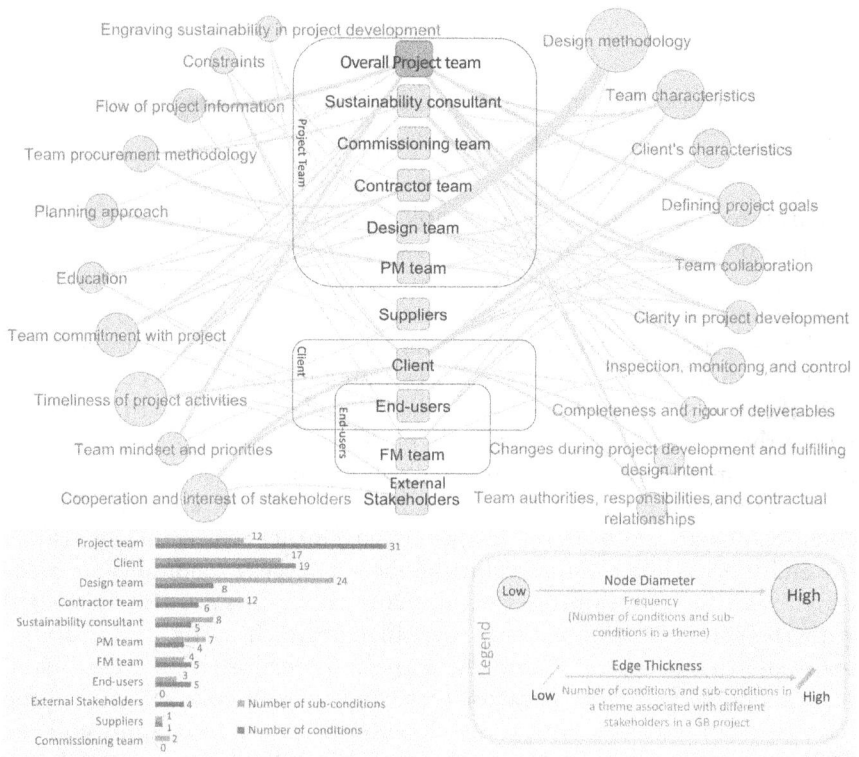

Figure 5.2 Associations of the themes of success conditions with Green Building project stakeholders.

Table 5.1 Associations of success conditions with Green Building project stakeholders

Theme	Sr. no.	Identified conditions	Overall participants	Overall project team	Design team	Contractor	Sustainability consultant team	PM team	Commissioning	Suppliers team	Client	FM team	End-users	External stakeholders
Changes during project development and fulfilling design intent	1	Change in project team members	2	X										
	2	Execution of sustainable design during construction	5				X				X			
	3	Scope changes during project execution	9											
	3.1	Permanence in client's requirements during project delivery	2								X			
	3.2	Client's motivation to achieve original project intent	3								X			
Clarity in project development	4	Clarity in the process of developing project	9	X							X			
	4.1	Use of clearly defined and standardised approaches for GB development	5	X							X			
	5	Delegating clear responsibilities to project team	6	X							X			
	5.1	Project team contractually required to deliver sustainable outcomes	2											
	5.2	Provision of sustainability specifications and other related information in tender	4			X								
Client's characteristics	5.3	Specificity of deliverables from Design Consultants	2		X									
	6	Client's leadership in project	3								X			
	7	Consensus within client organisation	2								X			
	8	Proficiency of project client	20								X			
	8.1	Client understanding the need of sustainable outcomes	5								X			
	8.2	Client's understanding of GB requirements	12								X			
	8.3	Client's understanding of sustainable building operation	2								X			
	8.4	Client's rational decision-making	1								X			
	9	Structure and nature of client organisation	5								X			
Completeness and rigour of deliverables	10	Completeness of project documentation for certification	2	X	X									
	11	Completeness and rigour of project design before execution	9		X	X								
	12	Setting of a detailed sustainability charter or brief	10								X			

(Continued)

Table 5.1 Associations of success conditions with Green Building project stakeholders (Continued)

Theme	Sr. no.	Identified conditions	Overall participants	Overall project team	Design team	Contractor team	Sustainability consultant	PM team	Commissioning team	Suppliers	Client	FM team	End-users	External stakeholders
Constraints	13	Access to sustainable building materials	2		X					X				
	14	Accessibility of project funding	6								X			
	15	Ease of logistics at project location	1										X	
Cooperation and interest of stakeholders	16	Client's involvement in project development	5								X			
	16.1	Client's facilitation of coordination among project team	1								X			
	17	Client's motivation to achieve sustainable outcomes	32								X			
	17.1	Client's endorsement of sustainability brief	3								X			
	17.2	Drivers for client to achieve sustainable outcomes	12								X			
	17.3	Investor's motivation to achieve sustainable outcomes	1								X			
	18	Cooperative role of building control authorities	2											X
	19	End-users' operation of building in sustainable ways	12										X	
	19.1	End-users' motivation to achieve sustainable outcomes	1										X	
	19.2	End-users' understanding of building operation	5										X	
	19.3	Tenants contractually required to consider sustainability	2										X	X
	20	Stakeholders' approval of project	4											X
Defining project goals	21	End-users' involvement in defining project aspirations	8										X	
	21.1	FM team involvement in defining project aspirations	1									X		
	22	Project team involvement in defining project aspirations	3	X										
	23	Setting appropriate project targets	21	X							X			
	23.1	Clarity in building's operational performance targets	1									X		
	23.2	Nature of GB certification aspired for project	2								X			
	23.3	Sustainability brief aligned with project budget	3					X						
	23.4	Specificity of project requirements	12	X										
	24	Stringency level of project sustainability requirements	4	X										
	24.1	Stringency level of GB certification requirements	2	X										

(Continued)

Table 5.1 Associations of success conditions with Green Building project stakeholders (Continued)

Theme	Sr. no.	Identified conditions	Overall participants	Overall project team	Design team	Contractor	Sustainability team	Sustainability consultant	PM team	Commissioning team	Suppliers team	Client	FM team	End-users	External stakeholders
Design methodology	25	Market survey to identify successful building systems	1									X			
	26	Rigour of project design development	41		X										
	26.01	Adding adaptability and multiple layers of use in building design	1		X										
	26.02	Use of reliable technology and solutions	2		X										
	26.03	Design consideration towards future building usage	3		X										
	26.04	Level of complexity in project design	4		X										
	26.05	Life-cycle-based project development approach	4		X										
	26.06	Maintainability considered in building design	3		X										
	26.07	Speculation in building design	7		X										
	26.08	Suitability of the project design for execution	3		X										
	26.09	Use of a proactive design approach	5		X										
	26.1	Use of a balanced design approach	1		X										
	26.11	Use of a context-oriented design approach	3		X										
	26.12	Use of an energy-oriented design approach	12		X										
	26.13	Use of a holistic design approach	9		X										
	26.14	Use of an innovative design approach	6		X										
	26.15	Use of an integrated design approach	13		X										
	27	Use of performance-based specifications	1		X	X									
Education	28	Educating client about sustainability in project	6									X			
	29	Educating end-users and FM team about building operation	9										X	X	
	30	Educating project team about GB development	8	X											
	30.1	Educating contractors about GB development	3			X									
	30.2	Educating sub-contractors about GB development	2			X							X	X	

(Continued)

Table 5.1 Associations of success conditions with Green Building project stakeholders (Continued)

Theme	Sr. no.	Identified conditions	Overall participants	Overall project	Design team	Contractor	Sustainability team	Sustainability consultant	PM team	Commissioning team	Suppliers team	Client	FM team	End-users	External stakeholders
Engraving sustainability in project development	31	Engraving sustainability in project development	8		X										
	32	Procurement of project site based on sustainability goals	3									X			
	33	Use of environmental management systems in construction	1			X									
Flow of project information	34	Clarity in communication of project goals	7	X											
	35	Communication among project team	11	X											
	35.1	Use of effective communication tools and strategies	5	X											
	36	Project team's access to robust information	5	X											
	36.1	Sharing of information related to project changes	1												
	37	Smooth transition of project from inception to operation stage	5	X									X		
Inspection, monitoring, and control	38	Inspection of project upon construction	8						X						
	38.1	Execution of commissioning and fine-tuning	7							X					
	39	Monitoring and controlling operational performance of building	5											X		
	40	Monitoring of project development	12						X						
	40.1	Project reporting	2						X						
	41	Review of project design by sustainability consultant	2					X							
Planning approach	42	Thoroughness of value engineering exercise	3	X											
	43	Attention towards details	5	X											
	44	Rigour of project planning	19						X						
	44.1	Adequate budget allocation for project development	13						X						
	44.2	Adequate time allocation for commissioning and testing	1						X						
	44.3	Adequate time allocation for project development	11						X						
	45	Rigour of risk management	4												

(Continued)

Table 5.1 Associations of success conditions with Green Building project stakeholders (*Continued*)

Theme	Sr. no.	Identified conditions	Overall project participants	Overall project team	Design team	Contractor team	Sustainability consultant	PM team	Commissioning team	Suppliers team	Client	FM team	End-users	External stakeholders
Team authorities, responsibilities and contractual relationships	46	Control of project design by Design and Sustainability Consultant	2		X		X							
	47	Empowerment of sustainability consultant by client	3	X			X				X			
	48	Project team's involvement in decision-making	2	X										
	49	Using appropriate project delivery method	3	X							X	X		
	49.1	Contractual interrelationships between client and project team	2	X							X			
Team characteristics	50	Like-mindedness of project team members	2									X	X	
	51	Proficiency of FM team	3	X										
	52	Proficiency of project team	32	X										
	52.1	Leadership qualities among project team members	2											
	52.2	Proficiency of contractor	12			X								
	52.3	Proficiency of Design Consultant	12		X									
	52.4	Proficiency of MEP Consultant	6		X									
	52.5	Proficiency of PM team	4					X						
	52.6	Proficiency of sub-contractor	5			X								
	52.7	Proficiency of sustainability consultant	5				X							
	53	Project team's understanding of project goals and aspirations	3	X										
	54	Size of design team	2		X									
Team collaboration	55	Conflicts among project team	1	X										
	56	Project team collaboration	29	X			X				X			
	56.1	Liaison between design team and client	5		X		X							
	56.2	Liaison between FM team and project development team	2								X	X		
	56.3	Liaison between sustainability consultant and client	1				X							
	56.4	Liaison between sustainability consultant and contractor team	3			X	X							
	56.5	Liaison between sustainability consultant and design consultant	4		X		X							
	56.6	Liaison between sustainability consultant and project team	5	X			X							
	56.7	Willingness of project team to work together	4	X			X							

(*Continued*)

Table 5.1 Associations of success conditions with Green Building project stakeholders (Continued)

Theme	Sr. no.	Identified conditions	Overall participants	Overall project	Design team	Contractor team	Sustainability consultant	PM team	Commissioning team	Suppliers	Client	FM team	End-users	External stakeholders
Team commitment to the project	57	Alignment of team interest with project interest	9	×										
	57.1	Rewards for achieving performance targets	2	×										
	58	Contractor's proactive role in project development	2			×								
	59	FM team motivation to achieve sustainable outcomes	1									×		
	60	Project team motivation to achieve sustainable outcomes	20	×										
	60.1	Contractor team motivation to achieve sustainable outcomes	8			×								
	60.2	Design Consultant's motivation to achieve sustainable outcomes	10		×									
	60.3	MEP Consultant's motivation to achieve sustainable outcomes	3		×									
	60.4	PM team motivation to achieve sustainable outcomes	3					×						
	60.5	Sustainability Consultant's motivation to achieve sustainable outcomes	2				×							
Team mindset and priorities	61	Establishing and promoting synergies	2	×										
	62	Focus of sustainability consultant on project goals	1				×							
	63	Open-mindedness and flexibility of project team	3	×										
	64	Priority of sustainability in project development	9	×										
	65	Team working on project with innovative mindset	1	×										
	66	Team working on project with value management mindset	5	×							×			
Team procurement methodology	67	Involvement of sustainability consultant in contractor's selection	1	×		×	×							
	68	Preferences in project team selection	12	×										
	68.1	Long-term engagement of sustainability consultant	2				×							
	68.2	Preferences in consultant's engagement	5		×									
	68.3	Preferences in contractor's engagement	5			×								
	68.4	Pre-qualification of contractors and sub-contractors	1			×								
	68.5	Requirement for contractor to engage a sustainability advisor	1			×								

(Continued)

Table 5.1 Associations of success conditions with Green Building project stakeholders (*Continued*)

Theme	Sr. no.	Identified conditions	Overall participants	Overall project team	Design team	Contractor	Sustainability consultant	PM team	Commissioning team	Suppliers team	Client	FM team	End-users	External stakeholders
Timeliness of project activities	69	Early engagement of project team	30		X									
	69.1	Early engagement of commissioning professionals	3						X					
	69.2	Early engagement of contractor	6			X								
	69.3	Early engagement of design consultants	1		X									
	69.4	Early engagement of FM team	6									X		
	69.5	Early engagement of sub-contractor	1			X								
	69.6	Early engagement of suppliers	2							X				
	69.7	Early engagement of sustainability consultant	22				X							
	70	Early introduction of project targets	22											
	70.1	Client's early decision-making regarding sustainability goals	4								X			
	70.2	Early incorporation of sustainability in project	17								X			
	71	Timeliness of building approval	2											X
	72	Timeliness of feedback on sustainability documentation	1											X
	73	Timely submission of GB certification documentation for review	2	X										
		Sum of conditions		31	8	6	5	4	0	1	19	5	5	4
		Sum of sub-conditions		12	24	12	8	7	2	1	17	4	3	0
		Sum of conditions and sub-conditions		43	32	18	13	11	2	2	36	9	8	4

Note: 'Contractor team' includes both the head contractor team and the sub-contractors; 'Design team' includes the architectural design consultants and the MEP design consultants; 'External Stakeholders' include the GB certification organisations, neighbourhood community, and building control authorities.

client ($n = 36$; 23%). A notable number of conditions are also associated with end-users ($n = 8$; 5%) and FM team ($n = 9$; 6%). The least number of conditions is associated with external stakeholders ($n = 4$; 3%) and suppliers ($n = 2$; 1%).

The detailed analysis reveals that for some conditions, the stakeholders affecting a condition and the stakeholders affected by a success condition are not the same. For instance, 'Clarity in process of developing project' is a condition which affects the project team. However, the project client has more control over this condition. This points towards an added layer of complexity in success conditions since one condition may affect and be affected by multiple stakeholders, enabling the success condition to affect the interests of multiple parties.

5.3 Assigning ownership of success conditions to project stakeholders

For the implementation of effective GB development practices, the success conditions identified in this study may need to be allocated to project stakeholders. The influence of different stakeholders in a GB project can vary depending on the type of project (for example public/private, office, and residential projects). Considering the client as the party which engages a project team for GB procurement, the client usually has the most influence over the project. The influence of the project team in project development depends on the authority that the client delegates to project team members. Both for the theoretical and practical reasons, it is important to know which stakeholders are effective in implementing which success conditions in GBs.

Pathways for the implementation of success conditions in GB projects are shown in Figure 5.3. The list of success conditions resulting from this study are readily applicable to GB projects in the form of checklists, and the project client and/or PM team can monitor their implementation. However, to effectively enable success conditions, it may be necessary to hold several project stakeholders accountable. To delegate responsibilities in this regard, the tradition of using risk owner/s in risk management can be visited. In risk management, a risk owner is an accountable point of contact for a risk, who coordinates efforts to mitigate and manage the risk with various individuals owning parts of the risk (Lavanya & Malarvizhi, 2008). The practice of successful GB project development and operation can also benefit from introducing 'success owners', an analogy of 'risk owners'. Like the role of risk owners, there can be several success owners in a project (such as clients, design consultants, and contractor) who coordinate efforts with the rest of the team to enable the success conditions assigned to them.

The assignment of success conditions to project stakeholders can be achieved through further research in this area. Future studies can indicate the project stakeholders who are most effective in controlling project development for specific success conditions. In practice, a limitation regarding

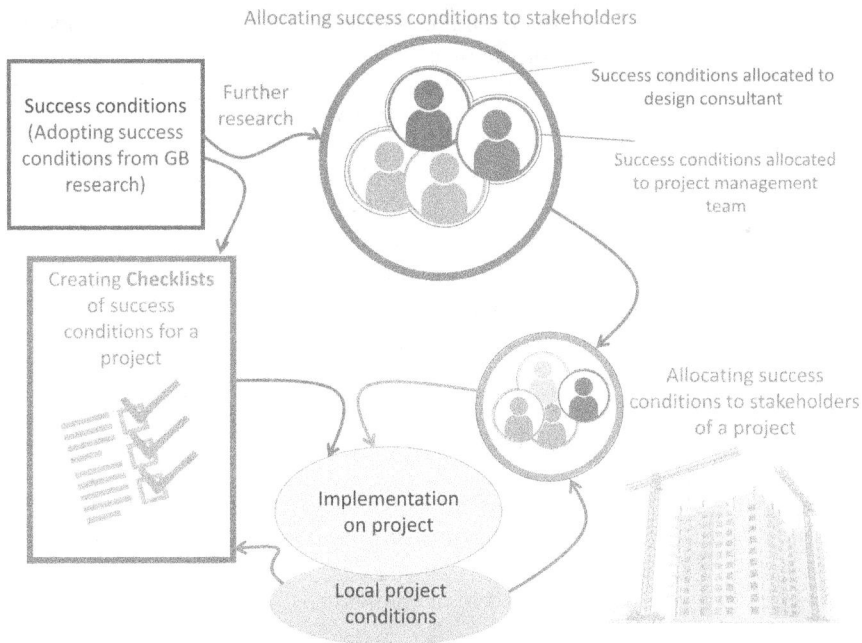

Allocating success conditions to stakeholders

Success conditions (Adopting success conditions from GB research)

Further research

Success conditions allocated to design consultant

Success conditions allocated to project management team

Creating **Checklists** of success conditions for a project

Allocating success conditions to stakeholders of a project

Implementation on project

Local project conditions

Figure 5.3 Implementation of Green Building success conditions by project stakeholders.

the allocation of success conditions to the project stakeholders is that they need to be adopted to the unique environment of each project. In future studies, allocating success conditions for success owners can be an important research inquiry for theoretical reasons. However, such allocations may need to be revisited before implementation and revised according to local project conditions, as indicated in Figure 5.3.

5.4 Suggestions related to Green Building stakeholders

The success conditions identified in this study are related to different GB stakeholders such as the client, end-users, and project team. These success conditions are presented in this section as suggestions for GB stakeholders.

5.4.1 Suggestions for project client

Based on the findings of this study, for ensuring project success, a GB project client is suggested to

Proficiency, motivation, leadership, and involvement

- Have the proficiency in project development particularly in terms of the understanding of GB requirements and sustainable building operation
- Provide leadership in project development

- Be motivated towards achievement of sustainable outcomes and endorse the sustainability brief of the project
- Get involved in project development and provide coordination among the project team

Setting goals

- Have clarity in communication of project goals
- Reduce the likelihood of scope changes during project execution
- Set detailed sustainability charter or brief
- Ensure the involvement of project team, end-users, and FM team in defining project aspirations
- Ensure that appropriate project targets are set with project requirements specifically defined and sustainability brief aligned with project budget
- Ensure that sustainability is engraved in project development
- Procure the project site based on sustainability goals

Ensuring rigour in planning and management

- Ensure rigorous project planning and in this regard allocate adequate time for project development
- Ensure rigour in project design development with careful speculation in building design and the use of integrated, energy-oriented, and life-cycle-based design approach
- Introduce project targets early, make decisions early in terms of sustainability goals, and incorporate sustainability early in the project
- Ensure that the end-users have the understanding of building operation and operate the building in sustainable ways
- Ensure that the neighbourhood communities and internal project stakeholders agree to project development
- Conduct rigorous risk management
- Ensure a rigorous inspection of the project upon construction and in this regard conduct commissioning and fine-tuning

Team procurement

- Engage project team early, particularly contractor, design consultant, and most importantly sustainability consultant
- Procure project team carefully, particularly the contractor and design consultant
- Ensure proficiency of project team for GB development, particularly the proficiency of contractor, design consultant, and PM team
- Empower sustainability consultants

5.4.2 Suggestions related to design consultant

Based on the findings of this study, for ensuring project success, a GB design consultant is suggested to

- Have the proficiency to develop GB projects
- Have the motivation to achieve sustainable outcomes
- Ensure that the design deliverables are well defined
- Ensure that the project design is complete and rigorously developed before execution
- Ensure that there is access to sustainable building materials
- Conduct a market survey for identifying successful building systems
- Ensure that performance-based specifications are used
- Engrave sustainability in project development
- Have a reasonable control of project design
- Liaise with sustainability consultant and client's team
- Ensure that the project design is rigorously developed
- Add adaptability and multiple layers of use in building design
- Use reliable technologies and solutions
- Ensure that during design development there is consideration towards future building usage
- Optimise the level of complexity in project design
- Use a life-cycle-based project development approach
- Ensure careful speculations for project design
- Consider building maintainability in design
- Consider the suitability of project design for execution
- Ensure that a proactive, balanced, context-oriented, energy-oriented, holistic, innovative, and integrated design approach is used

5.4.3 Suggestions related to contractor

The contractor of a GB project can contribute to project success, in case following aspects are considered during project development:

- Before project execution, a complete and rigorously developed project design is available
- Contractor team and sub-contractors are educated about GB development
- Performance-based specifications are used
- Contractor is required to engage a sustainability advisor
- Contractor and sub-contractors are engaged early in the project development
- There is the provision of sustainability specifications and other related information in project tender

- There is a pre-qualification process when engaging contractors and sub-contractors
- Contractor is carefully selected based on different criteria such as the ability to deliver GBs, perception of GBs, and previous experience of delivering GBs

For ensuring project success, the contractor of a GB project is suggested to

- Have the proficiency to develop GB projects
- Have the motivation to achieve sustainable outcomes
- Play a proactive role in project development
- Ensure rigorous and thorough execution of sustainable design during construction
- Use environmental management systems in construction
- Liaise with sustainability consultant

5.4.4 Suggestions related to sustainability consultant

The Sustainability consultant can contribute to GB project success, in case following aspects are considered during project development:

- Sustainability consultant is empowered by the client and has the control over project design
- Sustainability consultant is engaged early in project development and engaged for the entire duration of project development
- Sustainability consultant is involved in the contractor's selection

For ensuring GB project success, the sustainability consultant is suggested to

- Have the proficiency to develop GB projects
- Have the motivation to achieve sustainable outcomes
- Have the focus on project goals
- Review the project design in terms of sustainability considerations
- Liaise with project client, contractor team, design consultant, and other members of the project team

5.4.5 Suggestions related to project management team

For ensuring GB project success, the PM team is suggested to

- Have the proficiency to develop GB projects
- Have the motivation to achieve sustainable outcomes
- Ensure that project planning is rigorously conducted

- Ensure that an adequate budget is allocated for project development and the sustainability brief is aligned with project budget
- Ensure that adequate time is allocated for project development and particularly for commissioning and testing activities
- Ensure that risk management is rigorously conducted
- Monitor project development and conduct project reporting through different project stages to see if requirements are being met
- Inspect project upon construction

5.4.6 Suggestions related to facility management team

The FM team can contribute to GB project success, in case following aspects are considered:

- FM team is engaged early in project development and is involved in defining project aspirations
- The operational performance targets of building are clearly defined
- FM team is educated about building operation
- A smooth transition of the project from inception to operation stage is ensured particularly in terms of the transfer of building-related information

For ensuring GB project success, the FM team is suggested to

- Have the proficiency to work on GB projects
- Have the motivation to achieve sustainable outcomes
- Monitor and control the operational performance of the building
- Liaise with the project development team

5.4.7 Suggestions related to end-users

The end-users can contribute to GB project success, in case following aspects are considered:

- They are contractually required to consider sustainability.
- They are involved in defining project aspirations.
- They are educated about building operation.

For ensuring GB project success, the end-users are suggested to

- Have the motivation to achieve sustainable outcomes
- Have the understanding of building operation and operate the building in sustainable ways

5.4.8 Suggestions related to other project stakeholders

The likelihood of GB project success increases in case

- Commissioning professionals are engaged early in the project and they rigorously conduct commissioning and fine-tuning.
- Suppliers are engaged early in the project and they have access to sustainable building materials.
- Building control authorities cooperate in the GB project development. This can be in terms of providing timely approvals of project development and the provision of necessary infrastructure for meeting sustainability objectives in the project.
- Neighbourhood community and internal project stakeholders (that is client's organisation and end-users) approve the project development.
- GB certification organisation provides a timely feedback on the sustainability documentation of the project.

5.5 Summary

This chapter explored the role of project stakeholders regarding GB success conditions. A detailed account of the stakeholders responsible for and affected by GB success conditions is provided. There is a high association of success conditions with the overall project team, client organisation, and design team. Other stakeholders who affect or get affected by success conditions include sustainability consultant, PM team, commissioning team, contractor team, suppliers, end-users, FM team, and external stakeholders. For the practical application of success conditions, the idea of the ownership for success conditions in GB projects is discussed. The allocation of such ownership may require further research on the subject area.

Lastly, a detailed account of the suggestions for GB project stakeholders is provided to ensure the effective application of success conditions in GB projects. These suggestions can be readily adopted by the project team and stakeholders to ensure the successful development of GB projects.

Reference

Lavanya, N., & Malarvizhi, T. (2008). *Risk analysis and management: A vital key to effective project management.* Paper presented at the PMI® Global Congress 2008—Asia Pacific, Sydney, Australia.

Chapter 6

Annex A

Research design

6.1 Introduction

> **Who should read this chapter**
>
> Read this chapter:
>
> - If you want to know how the study questions about conditions differentiating Green Buildings (GBs) from non-GBs (that is, differentiating conditions) and conditions enabling success in GBs (that is, success conditions) are addressed
> - If you are interested in knowing why the findings in this book are based on semi-structured interviews of GB experts
>
> If you are interested in understanding differentiating conditions of GBs, see Part I. If you are interested in understanding success conditions of GBs, see Part II.

The research design determines the amount of control a researcher has over the research environment and guides what to observe and how to analyse the data (Inkoom, 2012; O'Sullivan, Berner, Taliaferro, & Rassel, 2016). This chapter provides a detailed overview of the research design, specifies precisely what is studied, and indicates the best approach to answer the research questions. In this chapter, the research paradigm, ontology, and epistemology used for this study are explained. A detailed discussion on semi-structured interviews is provided, followed by an explanation of the analysis and interpretation approaches used for interview data. Finally, a detailed account of research rigour is provided.

6.2 Research paradigm, ontology, and epistemology

Critical realism, a research paradigm based on ontological realism and epistemic relativism, is best suited for inquiry on GB project success and is therefore used in this study. While neither the pure positivist nor the interpretive

DOI: 10.1201/9781003322740-8

paradigms are suitable for GB project success inquiry, critical realism which sits between positivism and interpretivism can address the research on GB project success. Archer, Decoteau, Gorski, Little, and Porpora (2016) highlighted, 'Critical realism situates itself as an alternative paradigm both to scientific forms of positivism concerned with regularities, regression-based variable models, and the quest for law-like forms; and also to the strong interpretive or postmodern turn which denied explanation in favour of interpretation, with a focus on hermeneutics and description at the cost of causation'.

This study adopts a critical realism paradigm and accepts that reality exists and operates independent of our awareness or knowledge of it. With this stance, the study considers that GB projects and their development practices exist independent of our knowledge of them; in other words, their existence is not socially constructed. While using realism as ontology, in epistemic terms, relativism is adopted, which is based on the belief that knowledge about reality is always socially, culturally, and historically situated. Epistemic relativism is based on the understanding that knowledge is articulated from various standpoints according to various influences and interests and is transformed by human activity. While knowledge is context dependent, the representations of the world are historical, perspectival, and fallible (Archer et al., 2016). The study accepts that there is no way of knowing the world except under historically transient descriptions. The knowledge of successful GB project development is bound to the limits of the data collection methods used. The study accepts that the factors and their identified interrelationships regarding successful GB development are historically situated. As the GB market will evolve over time, innovative ideas and practices will arise, which project teams will use to improve the project development processes (Korkmaz, Horman, & Riley, 2009). While the research findings proposed in this study may represent current GB projects, they may not completely represent future GB projects. Besides the historical context, the demographics of study participants also influence the study findings as this can affect the perspectives employed in understanding the knowledge.

Using the foundations of epistemic relativism, this study considers project success as context-specific and develops the concept of GB project success based on the qualitative viewpoint of GB professionals.

6.2.1 Contingent and subjectivist: The approaches of project success inquiry

Three approaches to project success research include the 'objectivist', 'contingent', and 'subjectivist' approach (Ika, 2009), and this study adopts the contingent and subjectivist approach for GB project success inquiry as these are consistent with epistemic relativism. The 'objectivist' approach which is a dominant approach in project success research represents success frameworks as universal tools for achieving project objectives. For instance, Project Management Body of Knowledge (PMBOK) which is a flagship publication by the Project Management Institute[1] is presented as a fundamental resource for effective project management in any industry (Rose, 2013). The contingent approach considers success framework as context-specific, and the subjectivist approach is focused on defining project success from a subjective perspective resulting from a qualitative viewpoint (Ika, 2009). Considering GB projects as significantly different from traditional construction and acknowledging the experience of building professionals on GBs as a key source of knowledge on GB project success, the contingent and subjectivist approaches have informed the research design of this study.

Although the concept of project success remains ambiguous and lacks consensus from literature, it has been accepted as context-specific (Ika, 2009), and this is the position taken by the contingent approach. Previous research conducted on project success has demonstrated the impracticality of developing an exhaustive and generalised list of success conditions and criteria, which are applicable to all types of projects (Ika, 2009). As projects are unique in their complexity and scope (Ika, 2009; Wateridge, 1998), the importance attached to different success conditions and criteria varies with project type (Müller & Turner, 2007). In the case of GB project success, the contingent approach considers success in terms of the particular needs of sustainable development and operation of GBs. Dvir, Lipovetsky, Shenhar, and Tishler (1998) found the contingent approach as more suitable than identifying universal success criteria and conditions. The importance of using the contingent approach lies in accepting that each project type is unique, so the success conditions and criteria for one project type may not be the same as for other project types. Hence, there is no 'one best way' for defining project success. The contingent approach focuses on operationalising the aspects of a particular project in its actual context (Ika, 2009). In this way, the contingent approach offers a more meticulous and exclusive understanding of project success. Since GB projects are highly context specific in their development and operation, the research on GB project success can benefit from the exclusivity offered by the contingent approach and is therefore used in this study.

According to Baker, Murphy, and Fisher (1997), project success cannot be considered as an absolute concept and instead occurs as the 'perceived success of a project'. Different stakeholders may perceive the outcomes of the same project in different ways (Wateridge, 1998). A 'subjectivist' approach incorporates the variety of stakeholder perceptions of project success, providing a more comprehensive rationalisation of project success. An important aspect of this approach is the collection of stakeholder opinions from in-depth interviews, relying on the 'richness' of words instead of just preferences in terms of numbers. While the 'subjectivist' viewpoint is a break from mainstream research practices, it can open new avenues for the less investigated research areas of project success (Ika, 2009), such as GB project success, which is a research area in its early stages of development. For these reasons, this study has used the qualitatively driven subjectivist approach alongside the contingent approach.

By adopting epistemic relativism, this study acknowledges that our knowledge about reality is always socially, culturally, and historically situated. The context-specific contingent approach and the subjective perspectives of the subjectivist approach are synergetic with epistemic relativism and critical realism and are used in this study.

6.3 Semi-structured interviews

The face-to-face interview method is used, as it reduces non-response and maximises the quality of the data collected (Lavrakas, 2008). Semi-structured interviews, mainly consisting of themes, provide the opportunity to probe in detail some particular areas of interest (Lavrakas, 2008). In this study, the researcher engaged with the GB experts (that is, interview participants) in detailed discussions regarding the success conditions which were most relevant to their knowledge and experience. A key advantage of face-to-face interviews is the presence of the interviewer to clarify any issue that the respondents might face (Lavrakas, 2008). The researcher used a set of prompts to help the struggling respondents to provide distinct accounts of success conditions.

For open-ended in-depth questions, interviews provide more efficient access to information and have more potential to collect valid and reliable data. Although interviews are more time-consuming and expensive to conduct compared to questionnaire surveys, they are superior in terms of the access to comprehensive and detailed expert knowledge and also in terms of the ease of drawing implicit information (Akbayrak, 2000). These aspects of interviews were particularly relevant for this study, in identifying success conditions and in understanding their interrelationships. This is due to the fact that a lot of implicit information had to be made explicit. While potential bias exists in both survey questionnaire and interviews due to self-reporting (Akbayrak, 2000; Oppenheim, 2000), the bias in interviews was reduced by advanced training of the interviewer regarding the potential issues, careful articulation of questions, and by matching the attributes of the interviewer with the sample to be interviewed. Although higher anonymity and confidentiality is offered by questionnaires as compared to interviews (Akbayrak, 2000), these aspects did not become the decision-making criteria, as the discussion in interviews was not of a personal nature. Based on these considerations, the in-depth interview technique was selected as the data collection approach.

Some procedures were adopted to ease the access to information, reduce bias in findings, and ensure the quality of the collected data. These include the use of digital recording, the role of the researcher as both the interviewer and the transcriber, accurate transcription of interview discourse, the review of interviewee transcripts, and the use of computer-aided thematic coding for data analysis. The following sections provide a detailed overview of these aspects. Moreover, a detailed account of the themes in the interviews, the sampling approach, the demographics of participants, and the role of the researcher as the research instrument is also provided.

6.3.1 Themes in interviews

The themes in semi-structured interviews addressed the research questions posed in this study. Details regarding the themes used in interviews are as follows:

Differentiating conditions: To address the research question, that is, *How do sustainability requirements set Green Buildings apart from non-Green Buildings?*, the theme of differentiating conditions was used in interviews. Participants were requested to identify and elaborate on the aspects which set GBs apart from their traditional counterparts. In some cases, the participants reflected on these differentiating conditions while responding to some other questions related to GB development.

Success conditions: To address the research question, that is, 'What are Success conditions in Green Building projects and how do they interrelate?', the theme of success conditions was used in interviews. Three viewpoints are used to identify success conditions (as shown in Table 6.1), since this approach results in rigorous data collection and analysis. The viewpoint of *Findings from participants' overall experience* was intended to highlight success conditions for GBs, the interview participants had experienced in their overall professional careers of developing such projects. *Suggestions for clients* provided success conditions in control of the client developing the project. *Findings from successful/failed projects* highlighted the success and failure conditions of projects occurring in the careers of participants.

Table 6.1 Interview themes regarding success conditions and resulting viewpoints

No.	Interview theme	Resulting viewpoint
1	The conditions that are associated with success and failure of GB projects. The conditions occurring at the project development stage which can result in success/failure of a GB project during its operational life	Findings from participants' overall experience
2	Mention of some successful/failed GB projects in the interview participant's career. Categorisation of those projects as success/ failure. Conditions contributing to the success/failure of those projects. Conditions contributing to challenging circumstances in those projects. Lessons from those projects	Findings from successful/failed projects
3	Advice to a potential client of a GB project for successfully developing the project within limited resources	Suggestions for clients

While success examples are important to learn about the best practices of project development, the challenging and failed examples are also critical to learn about the conditions which enable success. For instance, Rasekh and McCarthy (2016) conducted a study to learn from the difficulties of delivering sustainable buildings. Considering the importance of both successful and failed projects, the viewpoint of *Findings from successful/failed projects* is used in conducting interviews. This viewpoint has an emphasis on learning from real-life projects and was also used by Hwang, Zhao, and Tan (2015) who engaged 34 participants in a questionnaire survey to investigate the schedule performance of 98 new and 51 retrofitted green building projects. Another seminal work in this regard is by Bond (2010) who investigated the best practices in GB development in Australia by conducting 23 interviews with GB stakeholders to collect their experiences related to 22 GB projects.

The use of multiple viewpoints for the success conditions theme helped identify a relatively large number of conditions per participant. By asking interview participants to use multiple viewpoints, the availability heuristic and bandwagon effect were avoided. The bandwagon effect is the tendency of believing what others believe (Henshel & Johnston, 1987), and the availability heuristic is a mental shortcut relying on immediate examples that come to a person's mind when evaluating a specific topic or concept. Under the availability heuristic, people tend to heavily weigh their judgements towards more recent information, making new opinions biased towards the latest news (Folkes, 1988). Another importance of using multiple viewpoints is that it results in the triangulation of findings and a detailed understanding of success conditions.

6.3.2 Administering interviews

While conducting interviews, the approach used in this study was to inquire the participants about aspects they are most suited to answer and cumulatively using this information to address the research questions. The study acknowledged that regarding a problem, knowledge can be comprised of many disjointed parts of information, possessed by one person or by many people. Normally, people do not realise how much they know about a problem since they are unaware of the information that may have relevance to a problem's solution. A very accurate estimate for a problem can be obtained by identifying relevant information (Alarcon-Cardenas & Ashley, 1992). Use of modularisation approach in conducting interviews can help acquire information from experts regarding independent modules of knowledge. For instance, successful design development and successful construction of a GB are two independent modules of inquiry. The design consultants can inform about conditions for successful design deliverables, while the contractors can inform the conditions which lead to successful building construction. Hence, GB experts with diversified roles in GB development can provide knowledge regarding different aspects of GB project success. Modularisation helps to ensure that concerning a subject, the experts' judgement and knowledge are accurate, and based on this, precise assessments can be made (Alarcon-Cardenas & Ashley, 1992). This study is mainly based on the information of GB project success acquired from the GB professionals. To effectively gather the relevant information from GB professionals in interviews, the modularisation approach has been used to some extent. The interview participants in this study had a variety of roles in GB projects. While following the modularisation approach, the researcher probed into those aspects of GB projects which were most relevant to an interview participant.

6.3.3 Sampling of interview participants

In this study, expert sampling was used to conduct interviews with industry professionals having experience of developing GB projects in Australia, Hong Kong, Pakistan, Singapore, the UAE, and the UK. Expert sampling which is a category of purposive sampling technique, calls for experts in a field to be the sampling subjects, and this sampling is particularly useful when investigating new areas of research (Etikan, Musa, & Alkassim, 2016). Owing to its merits for research on GB project success, the expert sampling approach is used in this study for conducting interviews. GB experts in Australia, Hong Kong, Pakistan, Singapore, the UAE, and the UK were contacted to participate in the interviews. While sampling for interviews, the contact details of experts were obtained from different online expert databases, including *GreenBookLive*,[2] *GBCA* database, and *USGBC* database. Other than Pakistan, all these regions are on the track of increased development of GBs, and these also have a significant number of professionals with extensive experience of developing GBs. Another reason for selecting different geographic regions was to consider the potential regional variations.

6.3.3.1 Using saturation to determine the sample size

When using purposive sampling for qualitative research, as in the case of this study, the sample size is often determined by data saturation, which is a state when the collection of new data does not shed any further light on the issue under investigation (Baker, Edwards, & Doidge, 2012; Mason, 2010; Morse, 2000). This study used a reasonably large sample size of 75 GB experts as interview participants, complying with the principle of data saturation.

6.3.3.2 Choice of regions in conducting interviews

As mentioned in Part II, interviews are conducted with GB professionals having experience of GB projects in Australia, Hong Kong, Singapore, the UAE, the UK, and Pakistan. The choice of regions for conducting interviews was based on the criteria of expert sampling, diversity of socio-economic conditions, and the feasibility for the researcher in terms of time and resources. The selected regions had building professionals with expertise in GB development.

These regions have a variety of drivers and challenges for GB market and have diverse socio-economic conditions to help develop comprehensive lists of factors (success conditions and differentiating conditions).

Different socio-economic conditions and the drivers and challenges of the GB market in different regions are also the reason behind the diverse regions covered. Based on the data for 2017, Australia had a very high Human Development Index (HDI) value of 0.939. In terms of the HDI value, Australia is followed by Hong Kong (HDI = 0.933), Singapore (HDI = 0.932), the UK (HDI = 0.922), and the UAE (HDI = 0.863). While these five regions have very high human development, Pakistan has a medium level of human development (HDI = 0.562) (United Nations Development Programme, 2018). A report by Petrullo, Jones, Morton, and Lorenz (2018), based on a global survey of more than 2,000 industry participants from 86 countries, analysed the GB market trends for many regions. For different regions, this report provides a detailed account of the triggers and challenges, as well as the social and environmental reasons for GB development. A detailed account of the GB market trends for the six regions (that is, Australia, the UAE, the UK, Singapore, Hong Kong, and Pakistan) is provided in Table 6.2.

Table 6.2 Drivers and barriers of Green Building development in six regions

		UK	UAE	Australia	Hong Kong	Singapore	Pakistan
Top challenges for GB development	Perception that green is for higher end projects only	x	x	x	x	x	
	Lack of political support/incentives					x	x
	Lack of trained/educated GB professionals			x		x	
	Lack of public awareness			x			x
	Lack of market demand	x					
	Lack of GB codes and regulations						x
Top triggers expected to drive future green activity in the region	Client demands	x			x	x	
	Market demands				x		
	Environmental regulations	x	x	x	x	x	
	Reduce operating costs					x	
Top social reasons for GB development	Improving occupant health and well-being	x	x	x	x	x	
	Promoting sustainable business practices	x	x	x	x	x	
	Increase worker productivity	x	x				
	Create a sense of community				x	x	
Top environmental reason for GB development	Reducing energy use	x	x	x	x	x	
	Reduce greenhouse gas emissions	x			x	x	
	Indoor air quality					x	
	Protect natural resources and water conservation				x		x

Sources: Drivers and barriers of GB development in the UK, the UAE, Singapore, Hong Kong, and Australia are obtained from the study conducted by Petrullo et al. (2018). Barriers of GB development in Pakistan are obtained from the study conducted by Azeem, Naeem, Waheed, and Thaheem (2017).

Alongside the similarities in triggers, barriers, and socio-economic reasons for GB development in the six regions, there are also differences in drivers and barriers, indicating differences in GB market trends across these regions. The difference in GB trends across the five regions is a function of the regional context. This highlights that GBs are developed to suit the social, economic, regulatory, and environmental conditions of a region. Therefore, conducting interviews in these regions can capture diverse GB development practices. As a mainly explorative inquiry, this study thrives on a diverse sample set whether it is in terms of the regional belonging of interview participants or in terms of their professional roles in GB projects (such as design consultants, contractors, clients, and end-users).

Green Building development in Australia

According to an Australia-based sustainability consultant (AU-M-19),

> A big movement in Australia regarding sustainable development came during the Sydney Olympics, when the sport facilities and other relevant venues were being developed. After the Sydney Olympics, in Australia and especially Sydney, the concept of sustainability became quite popular. At that time, we started to push the designs in terms of higher energy efficiency, and daylight use. This movement was driven by a lot of inspirational architects. This was an interesting time as there was a lot of re-learning that how the buildings used to be efficient in the past.

> Once the Green Star emerged, the industry as a whole was able to see what the green design was about. Afterwards, the terms like Green Star and ESD became a regular part of the vocabulary in industry. Green Star really pushed the supply chain from the construction perspective. The market matured quite quickly, and people got on-board as they saw in this a market advantage and a business potential. A lot of tier-1 contractors and developers embraced the concept of sustainability and sustainable design. There are still a lot of low tier contractors in the industry who don't fully embrace the concept of sustainability. However, some of the contractors have fully embraced it and it is part of their brand and product. Sustainable building development for Australia is quite mature these days in terms of professional skills. Culturally, sustainability is also well-integrated through a lot of contractors.

According to an Australia-based design consultant (AU-M-2),

> Australia has come late into this trend of sustainable development but there is a good chance that it will catch up with this trend. Australia in some strange ways is making some progressive moves toward sustainable development and bringing new stuff forward.

> In terms of Green Building-related legislation, Australia is certainly behind Europe. GB projects are also not a big concern here as Australia has a more temperate environment. It doesn't get the cold winters as in

Europe, and there is more concern towards solar gains than anything else. In line with these temperate conditions, there appears to be low incentive for sustainable development. However, there has been a gradual demand for Green Buildings with the arrival of Green Star system.

Green Building development in Singapore

According to a Singapore-based design consultant (SN-M-5),

In Singapore, the original Green Mark Master plan was very simple. It was about saving energy and water. But as it started to move on, it began to incorporate features which can ensure better living environment. Green Mark version 2015 was co-created by many professionals from different fields.

Things in Singapore are evolving in terms of Green Building development. We have started to look into real human needs ... The hardest part is about the People part of the Triple bottom line. We have slowly started to push the People in Triple Bottom Line on the top.

Singapore is very different from other regions. Here the government has very high expectations from projects. It somehow resembles the Confucius doctrine according to which the government should be comprised of scholars. In Singapore, the government is not only made up of political people, it also involves highly technical people. Because of this, the government becomes very systematic in planning things. A very important driver of Green Building development in Singapore is government policy.

According to a Singapore-based engineering consultant (SN-M-8),

About ten to twenty years ago, the Green Building development was mainly looking into the energy savings issue. This concept actually started from the developed regions such as the US and European countries. At that time, in the Asian region including China and Singapore the concept of Green Buildings was relatively new. Around then, in the Asian region, the developers were highly concerned about the higher and faster returns on investment. At that time, Singapore began to pay attention towards Green Buildings. It learnt from the experience of other countries and kept on improving. Singapore started Green Mark about 15 years ago and ever since we are improving the building standards. In Singapore, the awareness of Green Building development across the different relevant professionals was spread quite fast.

According to a Singapore-based engineering consultant (SN-M-9),

Within last ten years the Green Mark has matured a lot and throughout the years it has evolved from an educational level to its fifth version. Now, Green Building development in Singapore may not appear to be much incentive driven and it has become a norm as everyone is going for it. Now there are two main drivers playing an important role, the first driver is the regulation, and the second driver is the market.

Green Building development in Hong Kong

According to a Hong Kong-based design consultant (HK-M-5),

> The building stock in Hong Kong contains new as well as old build-
> ings ... In Hong Kong, about 90% of the energy use is for buildings.
> This accounts for about two-third of the Greenhouse gases. .... Years
> ago, the Overall Thermal Transfer Values for commercial buildings
> were established within the Hong Kong statutory requirements. Us-
> ing the statutory regulation of these values, the heat gain and energy-
> consumption in the buildings could be controlled. So about 20 years
> ago, in 1995 the energy consumption in commercial buildings was re-
> stricted to a certain allowable range. This, however, was not applicable
> on residential buildings and was only applied to commercial, hotel
> and office buildings. Around 10 years ago, Hong Kong started to get
> high-rise residential buildings with curtain walls having glass glazing.
> Typically, no control of heat gain was applied to those buildings. The
> problem was particularly serious for buildings facing the West, having
> a good harbour view but at the same time getting a lot of heat gain.
> These buildings became green houses, trapping heat within them. To
> deal with this scenario, the government developed some strategies and
> issued guidelines for the development of residential buildings which
> could reduce their energy consumption. So, Hong Kong first focussed
> on commercial buildings and then it focussed on residential buildings.
> Every five years, the benchmarks are reviewed, and the regulations be-
> come increasingly stringent.
>
> In Hong Kong we have sustainable building design guidelines devel-
> oped for all building types. The main objective is to improve urban en-
> vironment. For instance, improving urban ventilation at the street level.
> Since in Hong Kong the buildings are tightly packed together, they block
> the passage of wind resulting in a wall effect. Because of this, the ur-
> ban micro-climate is not good. The other big issue in Hong Kong is
> the Urban Heat Island effect because of a large number of buildings
> in a small area. These two issues were the motivating factors behind
> sustainable building design guidelines. Three main factors considered
> in these design guidelines include building permeability, building set-
> back, and greenery coverage. The history of this goes back to 2009 when
> Hong Kong government initiated an intensive public consultation with a
> focus on quality of space in Hong Kong. This consultation was conduct-
> ed by Council of Sustainable Development. The consultation resulted
> in more than 50 measures for improving the living environment. Some
> of the measures were included in sustainable building design guidelines.
> Another measure was to impose the development of Green Buildings in
> building planning regulations.
>
> Although the process of including Green Buildings in building plan-
> ning regulations started a long time ago, it is still incomplete. This can
> be related to the question that what is the urgent matter for the society
> at large. Right now, the issue is of land shortage. Since the land price

is high and the public wants to own a living, government is looking for more land to build on. In all this scenario, the Green Building is not the top priority. This acts as a regional challenge for sustainability. The last Chief Executive in Hong Kong was a surveyor, and at that time the prime focus was the number of accommodations instead of the quality of space.

Green Building development in the UK

According to a UK-based design consultant (UK-F-5),

UK is driven towards Green Building projects to reduce the climate decline. Regulations in UK are far strict. Clients, developers and tenants in the UK want GBs so there is both the commercial and regulatory demand for such buildings.

According to a UK-based sustainability consultant (UK-F-6),

In the UK there are two primary drivers of Green Buildings. Local authorities impose certain regulations that need to be fulfilled. Because of these regulations a second wave of demand is created among the convinced developers as Green Buildings help them improve their CSR credentials and through these projects they implement their sustainability agendas. There are two types of clients driven by these factors. One client type only wants to satisfy the local authorities in terms of Green performance and only thinks of meeting the bare minimum requirements .... On the contrary, is the case of clients who are under the convictions that a green development adds value to a project.

According to a UK-based sustainability consultant (UK-M-5),

In the UK, it is often the planning requirements and building regulations which drive the sustainability in building projects. Most projects in the UK need to be certified with a rating system and this is mandated by planning requirements. Particularly, building development in London is required to have a certain level of BREEAM certification.

Green Building development in the UAE

According to a UAE-based design consultant (AE-F-1),

Dubai follows LEED-based system and in Abu Dhabi they use Estidama .... In recent years, Dubai has set mandatory standards for green buildings which must be complied with, and these have a lot of over lapping with LEED.

According to a UAE-based sustainability manager (AE-M-3),

> Starting in 2008, the PCFC Trakhees in Dubai mandated the green de-
> velopment for areas under its jurisdiction and the building permits were
> only given if projects were developed in accordance with GB regulations.
> 25–30% of Dubai's land bank is under this jurisdiction. The rest of the
> area in Dubai is under Dubai Municipality which does not have a manda-
> tory GB rating system yet.

> There are three GB certification systems used in Dubai. One was adopted
> from LEED and the credit points were adjusted for the region .... For
> villas, Dubai developed its own regulations. This can be done without
> going to speciality consultants and can be performed by any competent
> architect. For warehouses, Dubai developed another set of regulations.

According to a UAE-based sustainability consultant (AE-F-2),

> It is important to see the aspects which drove the government for Green
> Building development. It appears that the government has realised
> that they cannot sustain the highly energy intensive development. The
> amount of energy buildings use is very high. Dubai has one of the biggest
> Carbon footprints in the world. Oil is going to run short eventually and
> the Government prefers to export their resources [oil] rather than having
> it consumed within the country for meeting the huge energy demands.
> In this regard, it appears that the leaders have realised that the trend of
> increasing energy demand is an important issue to address. Therefore,
> instead of the care for the environment, it is the preferences for reduced
> oil use in energy generation which is the major driver. Therefore, the
> driver is coming out of the necessity as the financial state of the UAE
> depends on the oil exports.

6.3.3.3 Demographics of interview participants

As compared to recommended guidelines, a significantly larger purposive sample of interview participants was used in this study as overall
75 participants participated in semi-structure interviews (see Table 6.3). In
interviews, 33% ($n = 25$) participants provided their response for differentiating conditions and 100% ($n = 75$) participants provided their response
for success conditions. This relatively large sample size is used because
interviews were conducted with slightly heterogeneous participants (that
is, GB-related stakeholders having various roles in GB projects and belonging to different regions). Interviews with subject matter experts were conducted in Australia ($n = 32$), Hong Kong ($n = 8$), Pakistan ($n = 2$), Singapore
($n = 14$), the UAE ($n = 6$), and the UK ($n = 13$).

The interview participants are 71% male and are highly experienced in
developing GBs, and 76% of them have worked in GBs as sustainability consultants, sustainability managers, or design consultants. A total of 69 out of

Table 6.3 Demographic details of interview participants and themes addressed by them

Participant ID	Success conditions	Differentiating conditions	Telephone	In-person	Region where experience based	Role in GB projects	Years of involvement in GBs
AE-F-1	X		X		UAE	Design consultant	10
AE-F-2	X			X	UAE	Sustainability consultant	7
AE-M-1	X	X		X	UAE	Sustainability consultant	3
AE-M-2	X	X		X	UAE	Sustainability consultant	10
AE-M-3	X	X	X		UAE	Sustainability manager with GB regulatory organisation	9
AE-M-4	X	X		X	UAE	Design consultant	12
AU-F-1	X	X	X		Australia	Sustainability manager with GB certification organisation	10
AU-F-10	X	X	X		Australia	Sustainability manager; design consultant	10
AU-F-11	X		X		Australia	Design consultant	12
AU-F-2	X			X	Australia	Design consultant	18
AU-F-3	X			X	Australia	Sustainability consultant	9
AU-F-4	X			X	Australia	Sustainability consultant	6
AU-F-5	X	X		X	Australia	Sustainability consultant	15
AU-F-6	X			X	Australia	Sustainability consultant	8
AU-F-7	X		X		Australia	Design consultant	10
AU-F-8	X		X		Australia	Design consultant	9
AU-F-9	X		X		Australia	Sustainability consultant	18
AU-M-1	X			X	Australia	Sustainability consultant	5
AU-M-10	X			X	Australia	Sustainability consultant	10
AU-M-11	X		X		Australia	Sustainability consultant	16
AU-M-12	X	X	X		Australia	Electrical engineer	10
AU-M-13	X			X	Australia	Sustainability consultant	11
AU-M-14	X		X		Australia	Sustainability consultant	15
AU-M-15	X		X		Australia	Sustainability manager	15

(Continued)

Table 6.3 Demographic details of interview participants and themes addressed by them (Continued)

Participant ID	Success conditions	Differentiating conditions	Telephone	In-person	Region where experience based	Role in GB projects	Years of involvement in GBs
AU-M-16	X	X	X		Australia	Head contractor	14
AU-M-17	X		X		Australia	Design consultant	20
AU-M-18	X		X		Australia	Sustainability consultant	35
AU-M-19	X		X		Australia	Sustainability consultant; engineering consultant	22
AU-M-2	X	X		X	Australia	Design consultant	8
AU-M-20	X		X		Australia	Sustainability and design manager with contractor	14
AU-M-21	X	X		X	Australia	Commercial specification consultant	5
AU-M-3	X			X	Australia	Engineering consultant; sustainability consultant	28
AU-M-4	X			X	Australia	Design consultant	10
AU-M-5	X	X	X		Australia	Design and BIM manager with contractor	10
AU-M-6	X			X	Australia	Design manager and director with Developer	15
AU-M-7	X	X		X	Australia	Sustainability consultant	7.5
AU-M-8	X			X	Australia	Project manager	15
AU-M-9	X		X		Australia	Quality and environmental manager with contractor	10
HK-F-1	X	X		X	Hong Kong	Sustainability consultant	2
HK-M-1	X			X	Hong Kong	Sustainability consultant	7
HK-M-2	X			X	Hong Kong	Sustainability consultant	7
HK-M-3	X			X	Hong Kong	Design consultant	10
HK-M-4	X			X	Hong Kong	Design consultant	10
HK-M-5	X			X	Hong Kong	Design consultant	10
HK-M-6	X			X	Hong Kong	Environmental and applications engineer with HVAC manufacturer	15
HK-M-7	X			X	Hong Kong	Design consultant	24
PK-M-1	X	X	X		Pakistan	Sustainability consultant; design consultant	3.5
PK-M-2	X	X	X		Pakistan	Lead engineer with contractor	1

(Continued)

Table 6.3 Demographic details of interview participants and themes addressed by them (Continued)

Participant ID	Success conditions	Differentiating conditions	Telephone	In-person	Region where experience based	Role in GB projects	Years of involvement in GBs
SN-F-1	×			×	Singapore	Building-users' representative	7
SN-F-2	×			×	Singapore	Sustainability consultant; design consultant	8
SN-M-1	×			×	Singapore	Engineering consultant (mechanical)	20
SN-M-10	×			×	Singapore	Engineering consultant; Sustainability manager	12
SN-M-11	×			×	Singapore	Design consultant	15
SN-M-12	×			×	Singapore	Energy manager	14
SN-M-2	×			×	Singapore	Commissioning service provider; sustainability manager	15
SN-M-3	×			×	Singapore	Facilities management professional	13
SN-M-4	×			×	Singapore	Facilities management professional	3
SN-M-5	×			×	Singapore	Design consultant	22
SN-M-6	×			×	Singapore	Sustainability consultant; engineering consultant (mechanical)	7
SN-M-7	×			×	Singapore	Engineering consultant (mechanical); sustainability manager	12
SN-M-8	×			×	Singapore	Engineering consultant (mechanical); sustainability manager	10
SN-M-9	×			×	Singapore	Engineering consultant (electrical); sustainability manager	10
UK-F-1	×			×	UK	Design consultant	10
UK-F-2	×	×		×	UK	Design consultant	11
UK-F-3	×	×		×	UK	Sustainability consultant	4
UK-F-4	×	×		×	UK	Sustainability consultant; environmental designer	5
UK-F-5	×	×		×	UK	Design consultant	15
UK-F-6	×	×		×	UK	Sustainability consultant	10
UK-M-1	×	×		×	UK	Sustainability consultant; design consultant	10
UK-M-2	×	×		×	UK	Sustainability consultant	11
UK-M-3 (Res)	×			×	UK	Sustainability consultant	7

(Continued)

Table 6.3 Demographic details of interview participants and themes addressed by them (Continued)

Participant ID	Success conditions	Differentiating conditions	Telephone	In-person	Region where experience based	Role in GB projects	Years of involvement in GBs
UK-M-4	x	x		x	UK	Energy auditor; engineering consultant	10
UK-M-5	x	x		x	UK	Sustainability consultant; design consultant	6
UK-M-6	x			x	UK	Design consultant	11
UK-M-7	x	x		x	UK	Design consultant; contractor	12
Number of participants	75	25	21	54			

Note: Key to participant ID: Part 1-Part 2-Part 3.

Part-1: Regional origins of the interview participant/region where the GB project experience of the interview participant is mainly based.

AE = UAE; AU = Australia; HK = Hong Kong; PK = Pakistan; SN = Singapore; UK = United Kingdom.

Part-2: The gender of interview participant.

M = Male; F = Female.

Part-3: Unique integer ID of an interview participant belonging to a pool of gender and region.

the overall 75 participants had an experience of five or more years related to GB projects. This supports the reliability and credibility of the interview findings. The participants had different roles in the development of GB projects, with the two most common roles among the participants being 'design consultant' ($n = 19$) and 'sustainability consultant' ($n = 22$). While the majority of interviews were conducted in-person (72%; $n = 54$), 28% ($n = 21$) were conducted over the telephone. Face-to-face interviews are the best choice when participants are geographically accessible to the researcher (Given, 2008), as it allows the researcher to have more control and access to rich information. In terms of telephone interviews, most were conducted with participants belonging to Australia ($n = 17$) as well as Pakistan ($n = 2$) and the UAE ($n = 2$).

6.3.4 Interview data analysis

The interview data is analysed using thematic analysis, which was used to classify the findings into groups and categories. Content analysis is used to quantitatively interpret the interview data and before conducting frequency analysis, creating factor networks, and performing network analysis.

6.3.4.1 Thematic analysis and coding

The thematic analysis consists of identifying, analysing, and reporting themes. The role of a theme is to capture important aspects of the data in relation to the research question and represent patterned response or meaning within the dataset. Considering the ability of thematic analysis to organise and describe the dataset (Braun & Clarke, 2006), it is used for a qualitative analysis of interview data. 'Theoretical' thematic analysis is an approach primarily driven by the researcher's theoretical or analytic interest in the area. Instead of a rich description of the overall data, this approach is more concerned with a detailed analysis of some aspects of the data and is suitable when research starts with a specific question (Braun & Clarke, 2006). Since this study started with specific research questions, theoretical thematic analysis was suitable for this study.

The thematic analysis in this study is conducted at two levels, that is, at the level of explicit meanings of data and at the level of underlying ideas in data. In the semantic approach of thematic analysis, the themes are identified within the explicit meanings of the data and the analyst is not looking for anything beyond what a participant has said or what has been written. For instance, when discussing success conditions, some interview participants ($n = 12$) mentioned that the client's understanding of GB requirements is important for GB project success. Using the semantic approach, the researcher considered this as a success condition. In latent approach, the analysis goes beyond the semantic content of the data and starts to identify or examine the underlying ideas, assumptions, conceptualisations, and ideologies (Braun & Clarke, 2006). For instance, some interview participants mentioned the client's understanding of GB requirements ($n = 12$), and the client's understanding of sustainable building operation ($n = 2$) as a success condition. Upon analysing these responses for underlying ideas, the researcher realised that the interview participants were mentioning the proficiency of the project client regarding GB development and operation. Based on the analysis of underlying ideas, proficiency of the project client was considered a parent success condition with multiple sub-conditions. Since both the levels of analysis are important for this study, thematic analysis in case of this study is conducted at the semantic level while identifying factors (that is, differentiating condition and success conditions) and also at the latent level while investigating hierarchies of factors.

The six steps of conducting thematic analysis as indicated by Braun and Clarke (2006) have been used in this study. The researcher in this study familiarised himself with the data by reading interview transcripts multiple times, and then the initial codes were generated, and themes were searched. The identified themes were reviewed by their cross-comparison. This was followed by defining and naming themes which finally lead to the development of the report.

In the thematic analysis, codes occur at a primary level and categories or themes occur at a secondary level. Themes are based on the analysis of codes rather than of data (Elliott, 2018). For this study, the interviews were manually transcribed by the researcher and then coded in a computer-based application, NVivo 12, which can systematically code large qualitative datasets (Wiltshier, 2011). Nvivo 12 also has the option of automatic coding of the data. Instead of using this technique, the manual coding technique was used throughout since it gives more control to the researcher. From the codes, themes emerged, which were further analysed and reported. Both the thematic analysis and coding process informed each other in a cyclic process. The coding for thematic analysis was also used in conducting the content analysis. Verbatim descriptions of participants' discourse are provided in this study. Participants are referred by unique IDs which can be used to refer to the information regarding the experience, professional role, and regional belonging of interview participants as shown in Table 6.3.

6.3.4.2 Classifying data into groups and categories

The construction of typologies and taxonomies, which are categories and groups within categories, is an important element of analyses. While using the data, categories, subgroups, and relationships among them should be established. Such categorisation reduces the number of potential variables, hence, making the data more manageable and easing the pattern detection and possible dependencies, or in other words 'causalities' (Fellows & Liu, 2015). The classification of GB success conditions into groups and categories is necessary for this study, and this categorisation not only eased the qualitative analysis but also the interpretation of findings. For instance, 20 broad themes represent 73 success conditions and 25 of these success conditions further represent 82 sub-conditions.

To reduce the potentially negative outcomes of data classification, it can be useful to provide a clear trail of classification exercise, as practised in this study. Data classification into groups and categories provides the ease of analysis and interpretation. However, this also carries the risk that some important variables are not well represented in groups, and as a result, the analysis and interpretations following the data classification may be flawed. Moreover, in a study involving a large number of factors, the analysis and interpretations may be limited to the level of groups and not the individual factors. This may not only reduce transparency but also the reproducibility of research. Acknowledging that the same factors may be classified differently with different viewpoints, the researcher in this study has provided a clear trail of data classification and has enabled transparency in the analysis. This can help future researchers use data from this study for comparison purpose to conduct the analysis on the same data with different perspectives, and even to use the findings of this study in longitudinal research.

6.3.4.3 Quantitative analysis of interview data

It is necessary to provide a visual representation of qualitative data as explained below. According to grounded theorists, it is essential for the theory-building process to create visual representations of the emerging theories (Charmaz, 2006; Clarke, Friese, & Washburn, 2017; A. L. Strauss, 1987; A. Strauss & Corbin, 1998; Verdinelli & Scagnoli, 2013). This qualitative tradition of inquiry strongly encourages the use of diagrams and figures to synthesise major theoretical concepts and their connections. For qualitative studies, the importance of visual displays lies at all stages of analysis. To suit the different requirements of data visualisation, network diagrams are used in this study.

For success conditions, network analysis is conducted in addition to the visualisation. This analysis is conducted using Cytoscape, a computer application used for visualising and analysing complex networks (Kohl, Wiese, & Warscheid, 2011).

Content analysis is a research technique used in making replicable and valid inferences by interpreting and coding textual material. Content analysis is used in this study to transform qualitative data into quantitative data. By systematically evaluating texts, the content analysis method offers the opportunity to convert qualitative data into numbers or percentages (Vaismoradi, Turunen, & Bondas, 2013; Xenarios & Tziritis, 2007) and enables a quantitative analysis of qualitative data (Wang, Wei, & Sun, 2013). The method has been applied in various areas of research on management, such as multi-criteria decision-making (Xenarios & Tziritis, 2007) and implementation of IS/IT strategy (Gottschalk, 2001). More so, it has also been used for GB project success research (Wang et al., 2013). Considering the usefulness of content analysis for this research inquiry, it has been adopted in addition to the thematic analysis.

In this study, content analysis has determined the frequency of occurrence of different factors, and these frequencies have determined the importance of identified factors. Whether to count codes or not is an important aspect to consider when reporting data. Researchers with pragmatic views consider counting important for a systematic approach to qualitative research. Counting may also serve as a useful indicator for the importance of a given code (Elliott, 2018). For instance, Harding (2018) suggests that a code shared by one-quarter of the participants in a study is worth consideration in the final analysis. Instead of the number of times a code appears in the data, its widespread occurrence in the data might be a more significant parameter to consider. For the quantitative analysis of qualitative research data obtained from interviews, the key parameter used in this study is 'frequency', which is based on the counting of codes. Frequency in this study means the number of participants identifying a factor (that is, success condition or differentiating condition) or highlighting an interrelationship among two factors. In a qualitative inquiry on GB project success, frequency has been used as a measure by Wang et al. (2013) to determine the relative importance of factors. Since the research data in case of this study is mostly qualitative, frequency analysis is used to interpret the relative importance of factors and to develop factor networks.

This study acknowledges that besides the usefulness of counting codes, there also exists the possibility that the frequency of occurrence does not necessarily indicate significance (Saldaña, 2015). Overreliance on counting may also result in a risk of overlooking significant and interesting concepts which have occurred in a dataset once or twice. While a code occurring at one, two, or three instances in a dataset may seem unimportant for a research question, it may be the key to unlock analysis (Saldaña, 2015). Hence the researcher has to be careful in recognising if that might be the case. If a researcher decides to count, it is essential that s/he also thinks carefully about how to use counts, and what they imply for the later stages of data analysis (Elliott, 2018). To avoid the drawbacks of overreliance on counting codes, the researcher in this study avoided the assumption that a factor is unimportant simply because it has low-frequency value.

6.3.5 Credibility of qualitative findings

Although the criteria used to evaluate qualitative research are not universal, some methodological strategies can enhance the credibility and trustworthiness of qualitative findings (Noble & Smith, 2015). These strategies used in this study include the following:

- Acknowledging biases in the sampling process and ensuring an ongoing critical reflection of the methods to achieve sufficient depth and relevance of data collection and analysis (Noble & Smith, 2015; Sandelowski, 1993). Furthermore, accounting for the personal biases which may have influenced the research findings (Morse, Barrett, Mayan, Olson, & Spiers, 2002; Noble & Smith, 2015).
- Meticulous record-keeping, demonstrating a clear decision trail and ensuring that the interpretations of data are consistent and transparent (Long & Johnson, 2000; Noble & Smith, 2015; Sandelowski, 1993). Also demonstrating clarity in the thought process during the analysis and interpretation of data (Noble & Smith, 2015; Sandelowski, 1993).
- Establishing a comparison case and seeking out the similarities and differences across accounts to ensure different perspectives are represented (Morse et al., 2002; Noble & Smith, 2015; Slevin & Sines, 1999). Also supporting the research findings by incorporating the rich verbatim descriptions of the participants' accounts (Noble & Smith, 2015; Slevin & Sines, 1999).
- Inviting the participants/respondents to comment on the interview transcripts and whether the final themes and concepts created by researcher adequately reflect the phenomena being investigated (Long & Johnson, 2000; Noble & Smith, 2015).
- Using data triangulation, whereby a more comprehensive set of findings is produced using different methods and perspectives (Fraser & Greenhalgh, 2001; Kuper, Lingard, & Levinson, 2008; Long & Johnson, 2000; Noble & Smith, 2015; Sandelowski, 1993).

Table 6.4 provides a detailed overview of how the above-mentioned methodological strategies are adopted in this study to enhance the credibility and trustworthiness of qualitative findings.

6.4 Summary

This chapter provides an overview of the research design used to address the research questions. The study adopts a critical realism paradigm, whereby a realist ontological position and a relativist epistemology are embraced.

While using the findings from semi-structured interviews, this study investigates the success conditions and differentiating conditions. Data

Table 6.4 Methodological strategies adopted to ensure the credibility of qualitative findings

Strategy	Use of strategy in this study
Accounting personal biases	An ongoing self-reflexivity of the researcher in the study to reduce the potential bias he may have introduced
Acknowledging biases in sampling and ongoing critical reflection of methods	Owing to the criteria used for expert sampling, some systematic bias in sampling may have been introduced While conducting interviews, the researcher practised caution, rigour, and self-reflexivity on an ongoing basis for effective data collection
Detailed record-keeping, having a clear decision trail and realising consistent and transparent interpretation of data	In terms of record-keeping, the interviews are audio-recorded and fully transcribed The interviews are coded in NVIVO application, which keeps an electronic record of the entire coding process. Besides its role in record-keeping, this approach also ensured consistency and transparency in data analysis and interpretation
Providing similarities and differences across accounts to represent different perspectives	During the thematic analysis, the researcher continuously compared the codes in interview transcripts Similarities and differences in the perspectives of interview participants are used while analysing and interpreting research findings
Verbatim descriptions of the participants' accounts	Verbatim descriptions are extensively provided while reporting the participants' accounts in this study. The verbatims are reported with the IDs of participants to help the readers see who has said something
Clarity in the thought process while analysing and interpreting data	A theoretical thematic analysis approach is used in this study to analyse the findings based on the preselected theoretical constructs
Validation by respondents	In this study, the Interviewee Transcript Review approach is used, whereby the interview participants are requested to review the transcripts
Data triangulation	Different approaches are used in this study to help data triangulation: A sampling criterion for interview participants is such that they cover a wide range of demographic settings. Hence, findings are triangulated from different demographic contexts The interviews also covered a variety of questions to help triangulate the findings. For instance, the interview participants were requested to discuss success conditions from three different viewpoints

is collected using semi-structured face-to-face interviews of GB experts. Procedures are adopted to ease the access to information, reduce bias in findings, and ensure the quality of collected data. These include the use of digital recording, the role of the researcher as interviewer and transcriber, knowledge acquisition by modularisation, the use of a denaturalised transcription approach, the use of interviewee transcript review process, and the use of computer-aided thematic coding for data analysis. While using the expert sampling approach for semi-structured interviews, saturation is used as the principle measure to determine the sample size. Overall, 75 GB professionals are interviewed belonging to Australia, Hong Kong, Singapore, the UAE, the UK, and Pakistan. Most of the participants ($n = 69$; 92%) had an experience of five or more years related to GB projects, and the majority of the participants ($n = 41$; 55%) have worked as either design consultants or sustainability consultants on GB projects. The multi-regional and professionally diverse interview participants have enabled diverse perspectives in data collection, which is important in terms of the explorative nature of this study.

To interpret the data and test the reliability of findings, qualitative and quantitative analysis are conducted on interview findings. The interview findings are analysed using a theoretical thematic analysis approach. The interview data is extensively classified into groups and categories for analysis and interpretation purpose. The interview findings are also quantitatively analysed, in terms of the frequency of identified factors and by network analysis.

To ensure the credibility and trustworthiness of qualitative findings, a number of methodological strategies are adopted, including the accounting of personal biases, accounting of biases in sampling, ongoing critical reflection of methods, detailed record-keeping, having clear decision trail, consistent and transparent interpretation of data, providing similarities and differences across the participants' accounts, clarity in the thought process in data analysis and interpretation, validation of transcripts by respondents, and data triangulation.

Notes

1   Project Management Institute is a global non-profit professional organisation for project management.
2   GreenBookLive is a database of Building Research Establishment Environmental Assessment Method (BREEAM), UK.

References

Akbayrak, B. (2000). A comparison of two data collecting methods: Interviews and questionnaires. *Hacettepe Üniversitesi Eğitim Fakültesi Dergisi, 18*(18). Retrieved from http://www.ijetmas.com/admin/resources/project/paper/f201604021459584494.pdf

Alarcon-Cardenas, L. F., & Ashley, D. B. (1992). *Project performance modeling: A methodology for evaluating project execution strategies* (Vol. 80). Austin, TX: Construction Industry Institute, University of Texas at Austin.

Archer, M., Decoteau, C., Gorski, P., Little, D., & Porpora, D. (2016). *What is critical realism?* Retrieved from http://www.asatheory.org/current-newsletter-online/what-is-critical-realism

Azeem, S., Naeem, M. A., Waheed, A., & Thaheem, M. J. (2017). Examining barriers and measures to promote the adoption of green building practices in Pakistan. *Smart and Sustainable Built Environment, 6*(3), 86–100. doi:10.1108/Sasbe-06-2017-0023

Baker, B. N., Murphy, D. C., & Fisher, D. (1997). Factors affecting project success. In D. I. Cleland & W. R. King (Eds.), *Project management handbook* (pp. 902–919). Hoboken, NJ: Wiley.

Baker, S. E., & Edwards, R., & Doidge, M. (2012). *How many qualitative interviews is enough? Expert voices and early career reflections on sampling and cases in qualitative research*. Retrieved from http://eprints.brighton.ac.uk/id/eprint/11632

Bond, S. (2010). Lessons from the leaders of green designed commercial buildings in Australia. *Pacific Rim Property Research Journal, 16*(3), 314–338. doi:10.1080/14445921.2010.11104307

Braun, V., & Clarke, V. (2006). Using thematic analysis in psychology. *Qualitative Research in Psychology, 3*(2), 77–101. doi:10.1191/1478088706qp063oa

Charmaz, K. (2006). *Constructing grounded theory: A practical guide through qualitative analysis*. Thousand Oaks, CA: Sage.

Clarke, A. E., Friese, C., & Washburn, R. S. (2017). *Situational analysis: Grounded theory after the interpretive turn*. Thousand Oaks, CA: Sage.

Dvir, D., Lipovetsky, S., Shenhar, A., & Tishler, A. (1998). In search of project classification: A non-universal approach to project success factors. *Research Policy, 27*(9), 915–935. doi:10.1016/S0048-7333(98)00085-7

Elliott, V. (2018). Thinking about the coding process in qualitative data analysis. *Qualitative Report, 23*(11), 2850–2861. Retrieved from https://nsuworks.nova.edu/tqr/

Etikan, I., Musa, S. A., & Alkassim, R. S. (2016). Comparison of convenience sampling and purposive sampling. *American Journal of Theoretical and Applied Statistics, 5*(1), 1–4. doi:10.11648/j.ajtas.20160501.11

Fellows, R. F., & Liu, A. M. (2015). *Research methods for construction*. Hoboken, New Jersey: Wiley.

Folkes, V. S. (1988). The availability heuristic and perceived risk. *Journal of Consumer Research, 15*(1), 13–23. doi:10.1086/209141

Fraser, S. W., & Greenhalgh, T. (2001). Coping with complexity: Educating for capability. *BMJ, 323*(7316), 799–803. doi:10.1136/bmj.323.7316.799

Given, L. M. (2008). *The Sage encyclopedia of qualitative research methods*. Thousand Oaks, CA: Sage.

Gottschalk, P. (2001). Descriptions of responsibility for implementation: A content analysis of strategic information systems/technology planning documents. *Technological Forecasting and Social Change, 68*(2), 207–221. doi:10.1016/S0040-1625(00)00084-6

Harding, J. (2018). *Qualitative data analysis: From start to finish*. Thousand Oaks, CA: Sage.

Henshel, R. L., & Johnston, W. (1987). The emergence of bandwagon effects: A theory. *The Sociological Quarterly, 28*(4), 493–511. doi:10.1111/j.1533-8525.1987.tb00308.x

Hwang, B. G., Zhao, X., & Tan, L. L. G. (2015). Green building projects: Schedule performance, influential factors and solutions. *Engineering, Construction and Architectural Management, 22*(3), 327–346. doi:10.1108/Ecam-07-2014-0095

Ika, L. A. (2009). Project success as a topic in project management journals. *Project Management Journal, 40*(4), 6–19. doi:10.1002/pmj.20137

Inkoom, E. E. (2012). *Sustainability adoption in construction organizations-an institutional and strategic choice perspective* (Master thesis). Retrieved from https://core.ac.uk/download/pdf/48659349.pdf

Kohl, M., Wiese, S., & Warscheid, B. (2011). Cytoscape: Software for visualization and analysis of biological networks. In *Data mining in proteomics* (pp. 291–303). New York, NY: Springer.

Korkmaz, S., Horman, M., & Riley, D. (2009). Key attributes of a longitudinal study of green project delivery. Paper presented at the Construction Research Congress, ASCE, Seattle, WA.

Kuper, A., Lingard, L., & Levinson, W. (2008). Critically appraising qualitative research. *BMJ, 337,* a1035. doi:10.1136/bmj.a1035

Lavrakas, P. J. (2008). *Encyclopedia of survey research methods.* Thousand Oaks, CA: Sage.

Long, T., & Johnson, M. (2000). Rigour, reliability and validity in qualitative research. *Clinical Effectiveness in Nursing, 4*(1), 30–37. doi:10.1054/cein.2000.0106

Mason, M. (2010). Sample size and saturation in PhD studies using qualitative interviews. *Forum qualitative Sozialforschung/Forum: Qualitative Social Research, 11*(3).

Morse, J. M. (2000). *Determining sample size.* Thousand Oaks, CA: Sage.

Morse, J. M., Barrett, M., Mayan, M., Olson, K., & Spiers, J. (2002). Verification strategies for establishing reliability and validity in qualitative research. *International Journal of Qualitative Methods, 1*(2), 13–22.

Müller, R., & Turner, R. (2007). The influence of project managers on project success criteria and project success by type of project. *European Management Journal, 25*(4), 298–309. doi:10.1016/j.emj.2007.06.003

Noble, H., & Smith, J. (2015). Issues of validity and reliability in qualitative research. *Evidence-Based Nursing, 18*(2), 34–35. doi:10.1136/eb-2015-102054

Oppenheim, A. N. (2000). *Questionnaire design, interviewing and attitude measurement.* Oppenheim, Germany: Bloomsbury Publishing.

O'Sullivan, E., Berner, M., Taliaferro, J. D., & Rassel, G. R. (2016). *Research methods for public administrators.* London: Routledge.

Petrullo, M., Jones, S., Morton, B., & Lorenz, A. (2018). World green building trends 2018–Smart market report. *Dodge Data & Analytics.* Retrieved from https://www.worldgbc.org/sites/default/files/World%20Green%20Building%20Trends%202018%20SMR%20FINAL%2010-11.pdf

Rasekh, H., & McCarthy, T. J. (2016). Delivering sustainable building projects–challenges, reality and success. *Journal of Green Building, 11*(3), 143–161. doi:10.3992/jgb.11.3.143.1

Rose, K. H. (2013). A guide to the project management body of knowledge (PMBOK® guide)—Fifth edition. *Project Management Journal, 44*(3), e1. doi:10.1002/pmj.21345

Saldaña, J. (2015). *The coding manual for qualitative researchers.* Thousand Oaks, CA: Sage.

Sandelowski, M. (1993). Rigor or rigor mortis: The problem of rigor in qualitative research. *Advances in Nursing Science, 16*(2), 1–8. Retrieved from https://www.ncbi.nlm.nih.gov/pubmed/8311428

Slevin, E., & Sines, D. (1999). Enhancing the truthfulness, consistency and transferability of a qualitative study: Utilising a manifold of approaches. *Nurse Researcher (through 2013), 7*(2), 79. doi:10.7748/nr2000.01.7.2.79.c6113

Strauss, A., & Corbin, J. (1998). *Basics of qualitative research techniques.* Thousand Oaks, CA: Sage.

Strauss, A. L. (1987). *Qualitative analysis for social scientists*. Cambridge: Cambridge University Press.

United Nations Development Programme. (2018). *Human development indices and indicators: 2018 statistical update*. South Africa: United Nations Development Programme.

Vaismoradi, M., Turunen, H., & Bondas, T. (2013). Content analysis and thematic analysis: Implications for conducting a qualitative descriptive study. *Nursing & Health Sciences*, *15*(3), 398–405. doi:10.1111/nhs.12048

Verdinelli, S., & Scagnoli, N. I. (2013). Data display in qualitative research. *International Journal of Qualitative Methods*, *12*(1), 359–381. doi:10.1177/160940691301200117

Wang, N., Wei, K., & Sun, H. (2013). Whole life project management approach to sustainability. *Journal of Management in Engineering*, *30*(2), 246–255.

Wateridge, J. (1998). How can IS/IT projects be measured for success? *International Journal of Project Management*, *16*(1), 59–63.

Wiltshier, F. (2011). Researching with NVivo. *Forum Qualitative Sozialforschung/ Forum: Qualitative Social Research*, *12*(1).

Xenarios, S., & Tziritis, I. (2007). Improving pluralism in multi criteria decision aid approach through focus group technique and content analysis. *Ecological Economics*, *62*(3-4), 692–703.

Chapter 7

Annex B

Exploratory analysis of success conditions

7.1 Introduction

> **Who should read this chapter**
>
> Read this chapter if you want to know how the differences in attributes and viewpoints of interview participants (GB experts) affect the identification of Green Building (GB) success conditions.
>
> If you are interested in understanding differentiating conditions of GBs, see Part I. If you are interested in understanding success conditions of GBs, see Part II.

To conduct this study on GB differentiating conditions and success conditions, epistemic relativism is adopted, which is based on the belief that knowledge about reality is always socially, culturally, and historically situated. Epistemic relativism indicates that knowledge is articulated from various standpoints according to various influences and interests and is transformed by human activity. For the inquiry of GB project success, the use of epistemic relativism implies that industry experts with differing roles and experiences in GB projects and different regional belonging may perceive the success in GBs differently. This chapter provides additional analysis to explore the effect of the attributes and viewpoints of interview participants on the identified success conditions.

7.2 Success conditions: Cross-comparison with viewpoints and attributes of interview participants

To analyse the success conditions against the viewpoints and attributes of interview participants, an itemised in-depth approach is used where each identified success condition is compared against each attribute and distinct viewpoint of the interview participant (see Table 7.1). The identified

DOI: 10.1201/9781003322740-9

success conditions and sub-conditions are shown in Table 7.1 which has 21 columns (A–T). The role of these columns in explaining the success conditions is as follows:

- Column A in Table 7.1 shows the overall number of interview participants ($n = 75$) who identified the conditions associated with GB project success. For instance, in the case of 'Project team's access to robust information', five participants had an opinion that this condition is important for GB project success. The conditions and sub-conditions associated with GB project success are identified by varying number of interview participants as shown in Figure 7.1. As shown, ten conditions and 15 sub-conditions are identified by only one participant in each case, while 14 conditions and 17 sub-conditions are identified by two participants in each case. Contrarily, ten conditions and four sub-conditions are identified by at least 13 participants in each case. In an extreme case, there is a success condition identified by 41 participants. The success conditions mentioned by a large number of interview participants include 'Rigour of project design development' ($n = 41$), 'Client's motivation to achieve sustainable outcomes' ($n = 32$), 'Proficiency of project team' ($n = 32$), 'Early engagement of project team' ($n = 30$), 'Project team collaboration' ($n = 29$), and 'Early introduction of project targets' ($n = 22$).

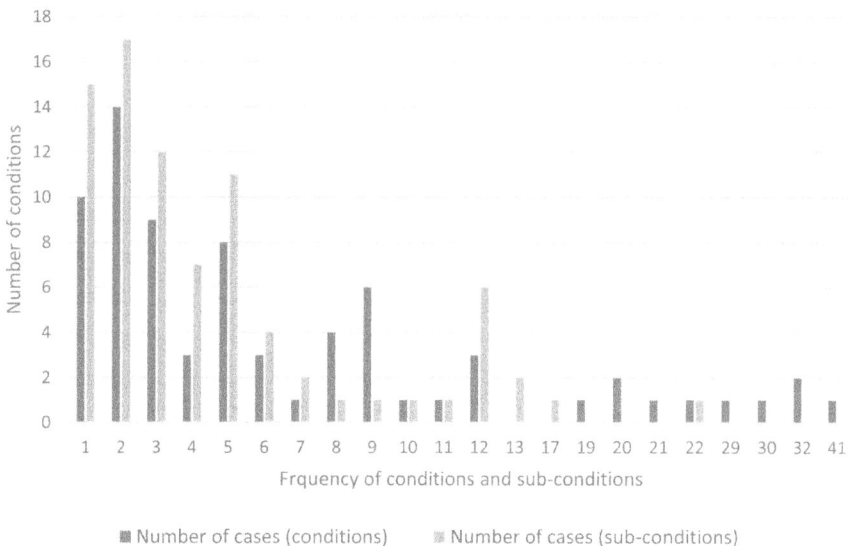

Figure 7.1 Frequency-based distribution of success conditions and sub-conditions.

Table 7.1 Identification of success conditions in different viewpoints by professionals belonging to six regions and having different experience of Green Building projects

Theme	Sr.no	Identified conditions and sub-conditions	A	B	C	D	E	F	G	H	U	I	J	K	L	M	N	O	P	Q	R	S	T
			Interviewees identifying success conditions	Conditions leading to challenges (case projects)	Failure conditions (case projects)	Failure conditions	Reasons for operational failures	Conditions behind the operational success	Success conditions (case projects)	Success conditions	Best practices in GB development	Suggestions for client	Design consultant	Sustainability consultant	UAE	Australia	UK	Hong Kong	Pakistan	Singapore	Low to medium (less than ten years)	High (10 or more years but less than 15 years)	Very high (15 years or more)
				Relation with failure in GBs				*Relation with success in GBs*			*Relation with GBs*		*Role in GBs*		*The region where participants' GB experience is based*						*Experience in GBs*		
Changes during project development and fulfilling design intent (n = 14)	1	Change in project team members	2	—	—	—	—	—	—	—	1	—	—	—	—	—	—	—	—	—	—	—	—
	2	Execution of sustainable design during construction	5	—	—	—	1	—	—	3	—	—	—	2	—	—	—	—	—	—	—	2	2
	3	Scope changes during project execution	9	—	3	3	—	—	—	2	—	—	3	2	2	3	2	2	1	1	3	2	4
	3.1	Permanence in client's requirements during project delivery	2	—	1	—	—	—	—	—	—	—	—	—	—	3	—	2	—	—	2	—	—
	3.2	Client's motivation to achieve original project intent	3	—	—	2	—	—	—	—	—	—	—	—	—	—	—	2	—	—	—	6	—
Clarity in project development (n = 13)	4	Clarity in process of developing project	9	—	—	—	—	6	—	4	3	—	4	3	—	4	3	—	2	—	2	2	—
	4.1	Use of clearly defined and standardised approaches for GB development	5	—	—	—	—	4	—	3	3	1	3	2	—	3	3	—	—	—	2	2	—
	5	Delegating clear responsibilities to project team	6	—	—	1	—	2	3	—	—	—	2	1	—	2	1	—	—	1	2	4	—
	5.1	Project team contractually required to deliver sustainable outcomes	2	—	—	—	—	—	2	—	—	—	1	1	—	1	1	1	—	—	—	—	—
	5.2	Provision of sustainability specifications and other related information in tender	4	—	—	—	—	3	2	—	2	—	—	—	—	—	2	—	—	1	—	3	—
	5.3	Specificity of deliverables from design consultants	2	—	—	1	—	—	—	1	—	—	—	—	—	1	—	—	—	—	—	2	—

		Category / Item																							
Client's characteristics (n = 24)	6	Client's leadership in project	3						1				2		1				1	1	1			
	7	Consensus within client organisation	2	1	1						2				2					1	1			
	8	Proficiency of project client	20	2	3	1	3	8	2	4	5	6	2	8	6	1	3	8	7	5			
	8.1	Client understanding the need of sustainable outcomes	5				1	4			2	2	2	1			2	2					
	8.2	Client's understanding of GB requirements	12	1	3		1	4	3	2	2	4		5	4		2	5	4	3			
	8.3	Client's understanding of sustainable building operation	2						1		4				1		2	1	1				
	8.4	Client's rational decision-making	1	1						1						1							
	9	Structure and nature of client organisation	5	1			4				4			2	2		3			1			
Completeness and rigour of deliverables (n = 20)	10	Completeness of project documentation for certification	2	1								1								1			
	11	Completeness and rigour of project design before execution	9	3			2	4	1		1	2	3	2	2	1	2	5	5	2			
	12	Setting of a detailed sustainability charter or brief	10	1			1	7	2	4	5	5	3	1		1	2	3	3	5			
Constraints (n = 9)	13	Access to sustainable building materials	2	2												1							
	14	Accessibility of project funding	6	2	1			3			3	1		4	2			4		1			
	15	Ease of logistics at project location	1																				
Cooperation and interest of stakeholders (n = 42)	16	Client's involvement in project development	5	1			1	1	2	1	2	1	2	1	1		2	1	2	2			
	16.1	Client's facilitation of coordination among project team	1																				
	17	Client's motivation to achieve sustainable outcomes	32	5			20	12	3	11	9	9	3	11	9	4	2	3	10	14	7		
	17.1	Client's endorsement of sustainability brief	3					1	2	2	1	2	2	2			3	1	1	2	2		
	17.2	Drivers for client to achieve sustainable outcomes	12	3		3	9		2	2	4	4	2	2	5	2	2	5	5	6			
	17.3	Investor's motivation to achieve sustainable outcomes					1				1						2						
	18	Cooperative role of building control authorities	2				2								1		1						
	19	End-users' operation of building in sustainable ways	12	1		2	3	3	1	6		6	1	5	2	2	3	3	5	5	4		
	19.1	End-users' motivation to achieve sustainable outcomes	1																	1	1		
	19.2	End-users' understanding of building operation	5			2				3		3					2	1	2	2	2		
	19.3	Tenants contractually required to consider sustainability	2									1		2						1			
	20	Stakeholders' approval of project	4	1			2	1		2		1		2	1		1	2					

(Continued)

Table 7.1 Identification of success conditions in different viewpoints by professionals belonging to six regions and having different experience of Green Building projects (*Continued*)

			Relation with failure in GBs					*Relation with success in GBs*					*Role in GBs*		*The region where participants' GB experience is based*						*Experience in GBs*		
			Interviewees identifying success conditions	*Conditions leading to challenges (case projects)*	*Failure conditions (case projects)*	*Failure conditions*	*Reasons for operational failures*	*Conditions behind the operational success*	*Success conditions (case projects)*	*Success conditions*	*Best practices in GB development*	*Suggestions for client*	*Design consultant*	*Sustainability consultant*	*UAE*	*Australia*	*UK*	*Hong Kong*	*Pakistan*	*Singapore*	*Low to medium (less than ten years)*	*High (10 or more years but less than 15 years)*	*Very high (15 years or more)*
Theme	*Sr.no*	*Identified conditions and sub-conditions*	*A*	*B*	*C*	*D*	*E*	*F*	*G*	*H*	*U*	*I*	*J*	*K*	*L*	*M*	*N*	*O*	*P*	*Q*	*R*	*S*	*T*
Defining project goals (*n* = 31)	**21**	End-users' involvement in defining project aspirations	8					1	2	5	1	2	5				4	3		1	1	5	2
	21.1	FM team involvement in defining project aspirations	1									1	1				1					1	
	22	Project team involvement in defining project aspirations	3						2			1	1	2	1		1	1			1	2	
	23	Setting appropriate project targets	21	1	2	4	1		6	10		3	7	5	2	6	5	2		6	6	11	4
	23.1	Clarity in building's operational performance targets	1				1							1		1					1		
	23.2	Nature of GB certification aspired for project	2	1	1								1	1			2				1	1	
	23.3	Sustainability brief aligned with project budget	3			1						2	1	1	1		1			1	1	1	1
	23.4	Specificity of project requirements	12		1	3			1	8		1	3	4	1	3	3	1		4	4	7	1
	24	Stringency level of project sustainability requirements	4	3						1				2		2	1	1			1	2	1
	24.1	Stringency level of GB certification requirements	2	1						1						2						1	1

	No.		41	2	1	11	6	1	21	4	3	12	11	3	16	7	5	1	10	10	19	12
Design methodology (n = 41)	25	Market survey to identify successful building systems	1																1		1	
	26	Rigour of project design development	41	2	1	11	6	1	21	4	3	12	11	3	16	7	5	1	10	10	19	12
	26.01	Adding adaptability and multiple layers of use in building design	2																			
	26.02	Use of reliable technology and solutions	3			1	1	1	1						2				1			
	26.03	Design consideration towards future building usage	4	1				2					2		4	2			2	2		
	26.04	Level of complexity in project design	4		1			3	2		2			4	4					1	1	1
	26.05	Life-cycle-based project development approach	3				2								3			3	2		
	26.06	Maintainability considered in building design	3				2				2	3	3	3				2	3	3	3	
	26.07	Speculation in building design	7		4		4	1			1	2						2	3	3	1	
	26.08	Suitability of the project design for execution	3		2		2				3	3		3	2			2	3	3	3	2
	26.09	Use of a proactive design approach	5			1	4		1				1					2	2			
	26.1	Use of a balanced design approach	1																	1		
	26.11	Use of a context-oriented design approach	3			3				2	2		2		2			2			2	2
	26.12	Use of an energy-oriented design approach	12			4	4	8		4	4	2	4	4	4	3	2	3	2	6	4	4
	26.13	Use of a holistic design approach	9			2	2	5		2	2	1	2	4	4	2	1	1	2	2	3	3
	26.14	Use of an innovative design approach	6	1	1		4	1		3			4	4		1	1		3	3	3	3
	26.15	Use of an integrated design approach	13		3		5	4	1	4	1	5	3	2	3	3	2	3	3	3	7	3
	27	Use of a performance-based specifications	1				1													2		
Education (n = 18)	28	Educating client about sustainability in project	6				5			5	2	1	3	3	1		2	2	1	2	2	4
	29	Educating end-users and FM team about building operation	9		2	5	3			3	1	1	5	1			3	3	1	4	4	4
	30	Educating project team about GB development	8			1	8			8	3	3	5	1			5	2	3	2	2	3
	30.1	Educating contractors about GB development	3				3			3	1											
	30.2	Educating sub-contractors about GB development	2				2			2		1					2	2	1	1		
Engraving sustainability in project development (n = 9)	31	Engraving sustainability in project development	8		4		4		3	4	2	2	4	2		3			1	2	3	3
	32	Procurement of project site based on sustainability goals	3				1		2	1		1	2	1		1	2			2	2	1
	33	Use of environmental management systems in construction	1									1										
Flow of project information (n = 25)	34	Clarity in communication of project goals	7	1		1	3		1	4	1		2	3	1	2	1		2	6	1	
	35	Communication among project team	11		1	9	3		6	1		3	6	4		3	4		2	7	3	3
	35.1	Use of effective communication tools and strategies	5			3	3		4			4	2	1		2	1		1	2	2	3
	36	Project team's access to robust information	5	2	2		1				3		4		3	4				1	3	1
	36.1	Sharing of information related to project changes	5	1																		
	37	Smooth transition of project from inception to operation stage	5		2	2	1			1	1	1	2	1		2	2			4		1

(Continued)

Table 7.1 Identification of success conditions in different viewpoints by professionals belonging to six regions and having different experience of Green Building projects (Continued)

Theme	Sr.no	Identified conditions and sub-conditions	A Interviewees identifying success conditions	B Conditions leading to challenges (case projects)	C Failure conditions (case projects)	D Failure conditions	E Reasons for operational failures	F Conditions behind the operational success	G Success conditions (case projects)	H Success conditions	I Best practices in GB development	J Suggestions for client	K Design consultant	L Sustainability consultant	M UAE	N Australia	O UK	P Hong Kong	Q Pakistan	R Singapore	S Low to medium (less than ten years)	T High (10 or more years but less than 15 years)	U Very high (15 years or more)
Inspection, monitoring, and control (n = 25)	38	Inspection of project upon construction	8		1	1	1	2		4			3	3	4	4	1	1		2	1	4	3
	38.1	Execution of commissioning and fine-tuning	7		1	1	2	2		3			3	3	4	4	1			1	3	3	3
	39	Monitoring and controlling operational performance of building	5				2			3			2	1	4	5	3			2	2	1	2
	40	Monitoring of project development	12			2				10			5	2	5	2	1			1	9	1	
	40.1	Project reporting	2																				
	41	Review of project design by sustainability consultant	2						1	1			1	1		1					1		1
	42	Thoroughness of value engineering exercise	3	1					1		1	2	2				2			1	2		
Planning approach (n = 24)	43	Attention towards details	5		1				2	2		2	2	2		2	2			1	1	3	2
	44	Rigour of project planning	19	4	1	6			3	8	1	3	7	7	8	5	1			5	3	9	4
	44.1	Adequate budget allocation for project development	13	3		4			1	6		3	3	3	6	2	1		3	3	6	3	
	44.2	Adequate time allocation for commissioning and testing	1												1								
	44.3	Adequate time allocation for project development	11			2				5	1	1	4		3	4	1			2	2	6	3
	45	Rigour of risk management	4					2	2	2		3	3	2	2	2	1		2	2	2	2	

			A	B	C	D	E	F	G	H	I	J	K	L	M	N	O	P	Q	R	S	T
			Relation with failure in GBs					Relation with success in GBs				Role in GBs			The region where participants' GB experience is based						Experience in GBs		

No.	Factor	Frequencies (read across)
	Team authorities, responsibilities, and contractual relationships (n = 9)	
46	Control of project design by design and sustainability consultant	2, 2, 1, 1
47	Empowerment of sustainability consultant by client	3, 1, 2
48	Project team's involvement in decision-making	2, 1, 2, 1
49	Using appropriate project delivery method	3, 2, 2, 2, 1
49.1	Contractual interrelationships between client and project team	2, 2, 2, 2, 1
	Team characteristics (n = 36)	
50	Like-mindedness of project team members	2, 1, 2
51	Proficiency of FM team	3, 2, 3, 2, 2, 2
52	Proficiency of project team	32, 1, 7, 6, 21, 2, 3, 7, 11, 1, 13, 8, 3, 2, 5, 11, 12, 9
52.1	Leadership qualities among project team members	2, 2, 5, 11
52.2	Proficiency of contractor	12, 4, 8, 1, 3, 3, 6, 3, 1, 4, 3, 5
52.3	Proficiency of design consultant	12, 2, 6, 2, 3, 3, 5, 3, 1, 3, 6, 5
52.4	Proficiency of MEP consultant	6, 1, 2, 2, 2, 4, 2, 5, 2
52.5	Proficiency of PM team	4, 2, 2, 3, 2, 2, 2
52.6	Proficiency of sub-contractor	5, 2, 3, 3, 2
52.7	Proficiency of sustainability consultant	5, 1, 4, 2, 2, 4
53	Project team's understanding of project goals and aspirations	3, 3, 3, 1, 3
54	Size of design team	2, 2, 2
	Team collaboration (n = 30)	
55	Conflicts among project team	1, 1, 1, 1
56	Project team collaboration	29, 1, 13, 15, 4, 6, 10, 3, 7, 9, 3, 2, 5, 14, 12, 5
56.1	Liaison between design team and client	5, 2, 2, 1, 1, 2, 2, 2
56.2	Liaison between FM team and project development team	2, 2, 2, 1, 2
56.3	Liaison between sustainability consultant and client	1, 1, 2, 1
56.4	Liaison between sustainability consultant and contractor team	3, 2, 2, 2, 1
56.5	Liaison between sustainability consultant and design consultant	4, 1, 2, 3, 2, 1
56.6	Liaison between sustainability consultant and project team	5, 4, 3, 4, 1
56.7	Willingness of project team to work together	4, 3, 2, 2, 3, 1
	Team commitment to the project (n = 27)	
57	Alignment of team interest with project interest	9, 1, 7, 3, 2, 4, 3, 1, 2, 5, 2
57.1	Rewards for achieving performance targets	2, 1, 3, 1
58	Contractor's proactive role in project development	2, 1, 1, 1
59	FM team motivation to achieve sustainable outcomes	1, 1, 1
60	Project team motivation to achieve sustainable outcomes	20, 2, 8, 1, 6, 7, 10, 5, 1, 2, 6, 10, 4
60.1	Contractor team motivation to achieve sustainable outcomes	8, 1, 3, 4, 3, 2, 6, 1, 3, 3, 2
60.2	Design consultant's motivation to achieve sustainable outcomes	10, 1, 3, 5, 3, 4, 1, 6, 3, 7, 2
60.3	MEP consultant's motivation to achieve sustainable outcomes	3, 1, 2, 2, 2
60.4	PM team motivation to achieve sustainable outcomes	3, 1, 3, 2
60.5	Sustainability consultant's motivation to achieve sustainable outcomes	2, 1, 1

(Continued)

Table 7.1 Identification of success conditions in different viewpoints by professionals belonging to six regions and having different experience of Green Building projects (Continued)

Column key:

Relation with failure in GBs:
- A = Interviewees identifying success conditions
- B = Conditions leading to challenges (case projects)
- C = Failure conditions (case projects)
- D = Failure conditions
- E = Reasons for operational failures
- F = Conditions behind the operational success

Relation with success in GBs:
- G = Success conditions (case projects)
- H = Success conditions
- U = Best practices in GB development
- I = Suggestions for client

Role in GBs:
- J = Design consultant
- K = Sustainability consultant

The region where participants' GB experience is based:
- L = UAE
- M = Australia
- N = UK
- O = Hong Kong
- P = Pakistan
- Q = Singapore

Experience in GBs:
- R = Low to medium (less than ten years)
- S = High (10 or more years but less than 15 years)
- T = Very high (15 years or more)

Theme	Sr.no	Identified conditions and sub-conditions	A	B	C	D	E	F	G	H	U	I	J	K	L	M	N	O	P	Q	R	S	T
Team mindset and priorities (n = 18)	61	Establishing and promoting synergies	2						2														1
	62	Focus of sustainability consultant on project goals	1																				1
	63	Open-mindedness and flexibility of project team	3							3						3	3						2
	64	Priority of sustainability in project development	9	2	2	2			1	3			2	2		3	3			1	2	5	2
	65	Team working on project with innovative mindset	1							3							3						3
	66	Team working on project with value management mindset	5						1	3			2	2			3						
	67	Involvement of sustainability consultant in contractor's selection	12						3	7		2	4	2	1	3	4	1	1	2	4	6	2
Team procurement methodology (n = 13)	68	Preferences in project team selection	2	1						2			1			2							
	68.1	Long-term engagement of sustainability consultant	2							2				1		2							
	68.2	Preferences in consultant's engagement	5						2	2							3	1	3		3	2	2
	68.3	Preferences in contractor's engagement	5							3		2					3				3	2	2
	68.4	Pre-qualification of contractors and sub-contractors	1							1													
	68.5	Requirement for contractor to engage a sustainability advisor	1			1									1							1	

		Total	1	2	3	4	5	6	7	8	9	10	11	12	13	14	15	16	17	18	19	20
Timeliness of project activities (n = 39)																						
69	Early engagement of project team	30	1	2		2	3	14	7	9	4	12	3	9	6	5	1	6	10	14	6	
69.1	Early engagement of commissioning professionals	3					1	1										2	2	3		
69.2	Early engagement of contractor	6	1			3	2		3		2	3		1		2	2	2				
69.3	Early engagement of design consultants	1				1			1			1		1		1	1					
69.4	Early engagement of FM team	6	2		4			1		2		1		3	3	4	1					
69.5	Early engagement of sub-contractor	1				1									1							
69.6	Early engagement of suppliers	2			2									1	1	1						
69.7	Early engagement of sustainability consultant	22	2		2	11	2	9	3	12	3	7	5	3	4	8	10	4				
70	Early introduction of project targets	22	1	6		3	7	4	5	10	7	10	5	3	1	2	9	8	5			
70.1	Client's early decision-making regarding sustainability goals	4				1	3		2	2	1	2	1	1		1	2	1				
70.2	Early incorporation of sustainability in project	17		6		6	4	2	6	7	8	3	2	2	7	6	4					
71	Timeliness of building approval	2				1				1	1											
72	Timeliness of feedback on sustainability documentation	1									1											
73	Timely submission of GB certification documentation for review	2	1				1	1	2	1	2	1	1									
	Number of interview participants	75	15	9	30	10	8	32	55	21	19	19	21	5	30	13	8	2	14	24	32	18
	Number of conditions identified	73	19	18	27	10	13	44	52	19	23	53	53	25	62	50	32	14	39	54	61	55
	Number of sub-conditions identified	82	11	8	29	11	5	38	54	20	24	52	56	17	56	54	29	11	36	58	67	49
	Number of unique conditions identified											5	3	0	8	3	0	0	2	3	6	4
	Number of unique sub-conditions identified											6	7	2	11	8	2	1	1	6	9	5

- Interview participants have discussed the conditions by explaining their role for project success from three viewpoints.

  - *Findings from successful/failed projects*: The interview participants recalled some successful/failed GB projects they have been involved in and highlighted the conditions resulting in challenges (Column B), failure (Column C), and success (Column G) in those projects. For instance, two interview participants associated 'Project team's access to robust information' with project challenges (Column B), three participants associated 'Scope changes during project execution' with GB project failure (Column C), and one interview participant associated 'Client's motivation to achieve original project intent' with GB project success (Column G).
  - *Findings from participants' overall experience*: Based on their overall experience of sustainable project development, the interview participants highlighted the conditions associated with success (Column H and Column U) and failure (Column D) of GB project development as well as the operational success (Column F) and operational failure (Column E) of GB projects. The difference between success conditions (Column H) and best practices (Column U) is that the success conditions according to participants are necessary for achieving desired outcomes while the best practices represent the conditions which positively influence the project performance, but the project success does not necessarily depend on these conditions.
  - *Suggestions for clients*: As experts of GB projects, some interviewees provided *Suggestions for clients* to successfully develop GB projects (Column I). For instance, as shown in Column I, two interview participants suggested 'Client's involvement in project development'.

- Although the GB professionals interviewed in this study have many different roles in GB projects (Figure PII.1 in Part II), design consultant and sustainability consultant are the two major roles of interview participants. Accordingly, Column J and Column K in Table 7.1 respond to the success conditions identified by design consultants ($n = 19$) and sustainability consultants ($n = 21$), respectively. It needs to be highlighted that the professionals represented in Column J and Column K only had one key role in GB projects.
- GB professionals interviewed in this study had a varying experience of GBs in terms of the number of years. In Table 7.1, the interview participants are divided into three groups, that is participants with less than ten years' experience of GB projects (Column R, $n = 22$), participants with ten or more years but less than 15 years' experience of GB projects (Column S, $n = 32$), and participants with 15 or more years' experience of GB projects (Column T, $n = 18$).
- GB professionals interviewed in this study have worked on GB projects in six geographical regions. Columns L, M, N, O, P, and Q respond

to the conditions identified by interview participants belonging to the UAE (n = 5), Australia (n = 30), the UK (n = 13), Hong Kong (n = 8), Pakistan (n = 2), and Singapore (n = 14), respectively.

Table 7.2 shows the results of analysing the success conditions identified by participants belonging to only a particular region, professional role, and professional experience. Table 7.2 highlights the conditions which are only associated with project success or project failure only, but not both. The explicit association of conditions with one of the three viewpoints (that is, *Findings from successful/failed projects*, *Findings from participants' overall experience*, and 'Suggestions for clients') is also shown in Table 7.2. This analysis provides the foundations for an in-depth discussion of identified conditions.

7.3 Analysis of success conditions based on the attributes of interview participants

The identified success conditions are analysed based on multiple contexts, that is participants' role in GBs, region of belonging, and experience of working on GBs. The key highlights of the analysis are provided in this section.

The analysis of identified conditions with respect to participants' role in GBs shows that 11 (7%) of the overall conditions are identified by design consultants only (shown in Table 7.2). While two of these conditions are about design methodology, the remaining nine conditions are about defining project goals, team characteristics, team commitment to the project, team procurement methodology, project constraints, cooperation and interest of stakeholders, and engraving sustainability in project development. As many as ten (7%) of the overall conditions are exclusively identified by sustainability consultants. These conditions are about changes during project development and fulfilment of design intent, defining project goals, planning approach, team mindset and priorities, team collaboration, the flow of project information, cooperation and interest of stakeholders, and timeliness of project activities.

The analysis of success conditions with respect to the regional context of interview participants reveals that participants from each region have identified some conditions which are not mentioned by participants from other regions. Table 7.2 shows the success conditions which are unique in case of each regional group of participants. The majority of success conditions (that is, 76%) prevail in multiple regions. Overall, 13 conditions and 25 sub-conditions (25% of the overall conditions) are unique in terms of the regional context; in other words, they are identified by participants belonging to only one out of the six regions. This validates that the identification of success conditions is a region-specific inquiry. The highest number of unique conditions and sub-conditions is identified in the case of Australia and the UK.

Table 7.2 Explicitness of success conditions and sub-conditions with respect to the viewpoints and demographic attributes of interview participants

Sr.no	Identified conditions and sub-conditions	Overall participants	Findings from successful projects	Findings from participants' failed projects	Suggestions for clients overall experience	Condition leading to failure	Condition leading to success	Design Consultant	Sustainability consultant	UAE	Australia	UK	Hong Kong	Pakistan	Singapore	Low to medium (less than ten years)	High (10 or more years but less than 15 years)	Very high (15 years or more)
											The region where participants' GB experience mainly based					Experience in GB projects		
								Role in GBs										
1	Change in project team members	2					X											
2	Execution of sustainable design during construction	5		X														
3	Scope changes during project execution	9																
3.1	Permanence in client's requirements during project delivery	2				X			X							X		
3.2	Client's motivation to achieve original project intent	3																
4	Clarity in process of developing project	9																
4.1	Use of clearly defined and standardised approaches for GB development	5					X											
5	Delegating clear responsibilities to project team	6					X											
5.1	Project team contractually required to deliver sustainable outcomes	2					X											
5.2	Provision of sustainability specifications and other related information in tender	4			X													
5.3	Specificity of deliverables from design consultants	2					X										X	
6	Client's leadership in project	3																
7	Consensus within client organisation	2				X					X							
8	Proficiency of project client	20	X															
8.1	Client understanding the need of sustainable outcomes	5					X											
8.2	Client's understanding of GB requirements	12																
8.3	Client's understanding of sustainable building operation	2																
8.4	Client's rational decision-making	1	X			X					X							X

No.	Item	n							
9	Structure and nature of client organisation	5	×						
10	Completeness of project documentation for certification	2			×				
11	Completeness and rigour of project design before execution	9							
12	Setting of a detailed sustainability charter or brief	10	×						
13	Access to sustainable building materials	2			×				
14	Accessibility of project funding	2							
15	Ease of logistics at project location	6							
16	Client's involvement in project development	1	×			×			
16.1	Client's facilitation of coordination among project team	5	×						
17	Client's motivation to achieve sustainable outcomes	1		×		×	×	×	
17.1	Client's endorsement of sustainability brief	32			×				
17.2	Drivers for client to achieve sustainable outcomes	3							
17.3	Investor's motivation to achieve sustainable outcomes	12	×		×	×		×	×
18	Cooperative role of building control authorities	1	×		×	×			
19	End-users' operation of building in sustainable ways	2	×					×	×
19.1	End-users' motivation to achieve sustainable outcomes	12		×	×	×	×	×	
19.2	End-users' understanding of building operation	1			×				
19.3	Tenants contractually required to consider sustainability	5					×		
20	Stakeholders' approval of project	2							
21	End-users' involvement in defining project aspirations	4							
21.1	FM team involvement in defining project aspirations	8		×	×	×		×	
22	Project team involvement in defining project aspirations	1			×				
23	Setting appropriate project targets	3							
23.1	Clarity in building's operational performance targets	21	×	×	×	×	×	×	
23.2	Nature of GB certification aspired for project	1	×	×					
23.3	Sustainability brief aligned with project budget	2							
23.4	Specificity of project requirements	3				×			
24	Stringency level of project sustainability requirements	12				×			
24.1	Stringency level of GB certification requirements	4							
25	Market survey to identify successful building systems	2	×		×	×		×	×
26	Rigour of project design development	1	×	×	×	×	×	×	×
26.01	Adding adaptability and multiple layers of use in building design	41	×						
26.02	Use of reliable technology and solutions	1	×						
26.03	Design consideration towards future building usage	2	×						
26.04	Level of complexity in project design	3				×			

(Continued)

Table 7.2 Explicitness of success conditions and sub-conditions with respect to the viewpoints and demographic attributes of interview participants (Continued)

Sr.no	Identified conditions and sub-conditions	Overall participants	Findings from successful projects	Findings from participants' failed projects	Findings from participants' overall experience	Suggestions for clients	Condition leading to failure	Condition leading to success	Design Consultant	Sustainability consultant	UAE	Australia	UK	Hong Kong	Pakistan	Singapore	Low to medium (less than ten years)	High (10 or more years but less than 15 years)	Very high (15 years or more)
26.05	Life-cycle-based project development approach	4																	
26.06	Maintainability considered in building design	3			×							×							
26.07	Speculation in building design	7																	
26.08	Suitability of the project design for execution	3			×													×	
26.09	Use of a proactive design approach	5						×											
26.1	Use of a balanced design approach	1			×			×	×			×						×	
26.11	Use of a context-oriented design approach	3			×			×											
26.12	Use of an energy-oriented design approach	12						×											
26.13	Use of a holistic design approach	9																	
26.14	Use of an innovative design approach	6																	
26.15	Use of an integrated design approach	13																	
27	Use of performance-based specifications	1	×					×									×		
28	Educating client about sustainability in project	6			×			×									×		
29	Educating end-users and FM team about building operation	9			×											×			
30	Educating project team about GB development	8			×			×											
30.1	Educating contractors about GB development	3						×				×							
30.2	Educating sub-contractors about GB development	2			×			×											
31	Engraving sustainability in project development	8																	
32	Procurement of project site based on sustainability goals	3						×											

No.	Item	n							
33	Use of environmental management systems in construction	1	x			x			x
34	Clarity in communication of project goals	7	x	x			x		
35	Communication among project team	11	x			x			
35.1	Use of effective communication tools and strategies	5	x	x					
36	Project team's access to robust information	5							
36.1	Sharing of information related to project changes	1	x		x		x		
37	Smooth transition of project from inception to operation stage	5	x			x			
38	Inspection of project upon construction	8			x				
38.1	Execution of commissioning and fine-tuning	7							
39	Monitoring and controlling operational performance of building	5	x	x					
40	Monitoring of project development	12	x	x					
40.1	Project reporting	2	x	x		x			
41	Review of project design by sustainability consultant	2					x		x
42	Thoroughness of value engineering exercise	3			x				
43	Attention towards details	5							
44	Rigour of project planning	19							
44.1	Adequate budget allocation for project development	13							
44.2	Adequate time allocation for commissioning and testing	1	x	x		x			x
44.3	Adequate time allocation for project development	11		x		x			
45	Rigour of risk management	4		x					
46	Control of project design by design and sustainability consultant	2	x	x	x				
47	Empowerment of sustainability consultant by client	3		x	x				
48	Project team's involvement in decision-making	2	x						
49	Using appropriate project delivery method	3		x					
49.1	Contractual interrelationships between client and project team	2		x		x			
50	Like-mindedness of project team members	2	x	x		x x			
51	Proficiency of FM team	3	x						
52	Proficiency of project team	32							
52.1	Leadership qualities among project team members	2	x	x					
52.2	Proficiency of contractor	12							
52.3	Proficiency of design consultant	12		x					
52.4	Proficiency of MEP consultant	6							
52.5	Proficiency of PM team	4							
52.6	Proficiency of sub-contractor	5	x						
52.7	Proficiency of sustainability consultant	5	x	x					
53	Project team's understanding of project goals and aspirations	3	x	x					x

(Continued)

Table 7.2 Explicitness of success conditions and sub-conditions with respect to the viewpoints and demographic attributes of interview participants (Continued)

Sr.no	Identified conditions and sub-conditions	Overall participants	Findings from successful/unsuccessful projects	Findings from participants' failed projects	Suggestions for clients overall experience	Condition leading to failure	Condition leading to success	Role in GBs		The region where participants' GB experience mainly based						Experience in GB projects		
								Design Consultant	Sustainability consultant	UAE	Australia	UK	Hong Kong	Pakistan	Singapore	Low to medium (less than ten years)	High (10 or more years but less than 15 years)	Very high (15 years or more)
54	Size of design team	2	X			X										X		
55	Conflicts among project team	1	X			X					X						X	
56	Project team collaboration	29																
56.1	Liaison between design team and client	5					X											
56.2	Liaison between FM team and project development team	2		X			X									X		
56.3	Liaison between sustainability consultant and client	1	X				X		X				X			X		
56.4	Liaison between sustainability consultant and contractor team	3					X											
56.5	Liaison between sustainability consultant and design consultant	4																
56.6	Liaison between sustainability consultant and project team	5		X			X						X					
56.7	Willingness of project team to work together	4					X											
57	Alignment of team interest with project interest	9		X														
57.1	Rewards for achieving performance targets	2		X			X											
58	Contractor's proactive role in project development	2																
59	FM team motivation to achieve sustainable outcomes	1		X			X	X			X							X
60	Project team motivation to achieve sustainable outcomes	20																
60.1	Contractor team motivation to achieve sustainable outcomes	8																
60.2	Design consultant's motivation to achieve sustainable outcomes	10																
60.3	MEP consultant's motivation to achieve sustainable outcomes	3										X						
60.4	PM team motivation to achieve sustainable outcomes	3	X															

No.	Item														
60.5	Sustainability consultant's motivation to achieve sustainable outcomes	2			X										X
61	Establishing and promoting synergies	2			X										
62	Focus of sustainability consultant on project goals	1	X		X									X	
63	Open-mindedness and flexibility of project team	3	X		X										
64	Priority of sustainability in project development	9													
65	Team working on project with innovative mindset	1	X		X	X								X	
66	Team working on project with value management mindset	5			X	X									
67	Involvement of sustainability consultant in contractor's selection	1		X	X									X	
68	Preferences in project team selection	12													
68.1	Long-term engagement of sustainability consultant	2	X		X				X	X					
68.2	Preferences in consultant's engagement	5			X										
68.3	Preferences in contractor's engagement	5													
68.4	Pre-qualification of contractors and sub-contractors	1	X		X	X							X		
68.5	Requirement for contractor to engage a sustainability advisor	1			X	X	X						X	X	
69	Early engagement of project team	30													
69.1	Early engagement of commissioning professionals	3			X	X								X	
69.2	Early engagement of contractor	6													
69.3	Early engagement of design consultants	1							X	X				X	
69.4	Early engagement of FM team	6	X		X	X			X						
69.5	Early engagement of sub-contractor	1	X		X				X	X					
69.6	Early engagement of suppliers	2	X		X										
69.7	Early engagement of sustainability consultant	22													
70	Early introduction of project targets	22													
70.1	Client's early decision-making regarding sustainability goals	4		X	X										
70.2	Early incorporation of sustainability in project	17													
71	Timeliness of building approval	2			X										
72	Timeliness of feedback on sustainability documentation	1					X		X	X		X			
73	Timely submission of GB certification documentation for review	2										X	X		
	Sum of conditions	13	15	0	28	5	3	0	8	3	0	2	3	6	4
	Sum of sub-conditions	8	26	2	41	6	7	2	11	8	2	1	6	9	5
	Sum of conditions and sub-conditions	21	41	2	69	11	10	2	19	11	2	3	9	15	9

A significant correlation (R-squared value = 0.80) is found among the number of interview participants belonging to different regions and the number of conditions and sub-conditions identified by these interviewees as shown in Figure 7.2. The correlation has a steep slope value of 3.22 which implies that as the number of interview participants belonging to a particular region increases, the number of identified conditions also increases. For instance, 14 participants from Singapore identified 75 success conditions. With a notably larger number of participants from Australia ($n = 30$), a relatively larger number of success conditions (that is, 118) were identified. This indicates that the more number of interviews are conducted in a region, the more comprehensive list of success conditions is developed. This, however, does not mean that saturation is not possible in this case, that is the notion that with an increased number of participants there will always be new factors identified. The bar chart as shown in Figure 7.2

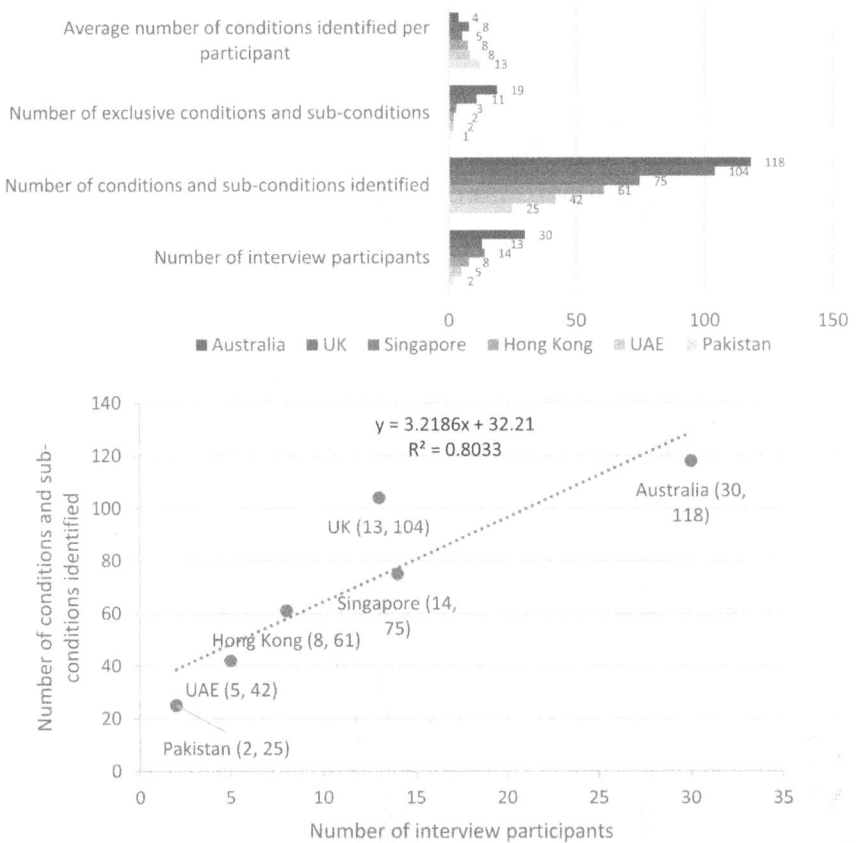

Figure 7.2 Correlation of the number of interview participants belonging to six regions with the number of identified conditions.

addresses the issue of saturation. As an example, in case of the UAE, 42 conditions were identified by five participants. This means that on average 8.4 conditions were identified by each participant. Contrary to this in case of Australia, 118 conditions were identified by 30 participants. This means an average of four conditions per participant. This shows a significant reduction in the average number of success conditions per participant with the increase in the number of participants. Moreover, 118 conditions in case of Australia do not imply that all those conditions were exclusively identified by Australian participants. Only 19 of those conditions were exclusively identified by Australian participants and not by participants from other regions. This highlights that even though the number of identified conditions is a function of sample size, saturation occurs in the identified conditions.

Analysis of success conditions with respect to the experience of participants in GBs reveals that each group of participants has identified some conditions and sub-conditions which are not mentioned by the other two groups. Table 7.2 shows the success conditions which are unique in the case of each group of participants. Some success conditions identified only by participants with low- to medium-level experience include 'Use of performance-based specifications', 'Size of design team', and 'Timeliness of feedback on sustainability documentation'. Success conditions identified only by participants with high-level experience include 'Ease of logistics at project location', 'Market survey to identify successful building systems', 'Project team's understanding of project goals and aspirations', 'Conflicts among project team', 'Focus of sustainability consultant on project goals', and 'Involvement of sustainability consultant in contractor's selection'. Success conditions identified only by participants with very high-level experience include 'Use of environmental management systems in construction', 'Review of project design by sustainability consultant', 'FM team motivation to achieve sustainable outcomes', and 'Team working on project with innovative mindset'.

Overall, 16 conditions and 30 sub-conditions are identified by participants belonging to single instead of multiple contexts (that is, the region, professional role, and experience in GB projects as shown in Table 7.2). This highlights that although 70% of conditions are identified as success conditions in multiple contexts, the remaining 30% of conditions (which is considerable) are exclusive to a single context, that is identified by design consultants or identified by the UK-based participants only. The identification of conditions within multiple contexts results in an internal validation of success conditions and implies that at least 70% of the identified conditions could be generalisable.

To visualise how the experience of interview participants and the viewpoint used by participants affect their perception of success conditions, the top ten success conditions with highest frequency values are plotted on radars. Following are the key highlights of the analysis presented in Figure 7.3:

- In Part 1, the radar is plotted with respect to participants' region of belonging. The participants accountable for the highest and lowest frequency of success conditions belong to Australia and Pakistan, respectively. This is because of the large number of participants from Australia ($n = 30$) and the scarce number of participants from Pakistan ($n = 2$). While Australia-based participants are not strongly convinced of 'Project team collaboration' as a success condition, the UK- and the UAE-based participants have strongly highlighted this condition. Moreover, unlike Australia-based participants, the Singapore-based interviewees have strongly highlighted 'Setting appropriate project targets' as a success condition.
- In Part 2, the radar is plotted with respect to participants' experience of GB projects. Some dissimilarities in trends are worth mentioning. For instance, as compared to participants with high to very high experience, the interviewees with low- to medium-level experience have strongly highlighted 'project team collaboration', 'early introduction of project targets' and 'proficiency of project client' as success conditions.
- In Part 3, the radar is plotted with respect to participants' role in GB projects. It can be observed that a similar number of design consultants and sustainability consultants interviewed have a significantly different focus on success conditions. Unlike design consultants, the sustainability consultants have strongly highlighted some conditions, including 'proficiency of project team', 'early engagement of project team', and 'project team collaboration'. On the contrary, 'Rigour of project planning' is strongly highlighted by design consultants and not by sustainability consultants.
- In Part 4, the radar is plotted in terms of the three viewpoints used in interviews for the inquiry of success conditions. Within the viewpoint of *Findings from participants' overall experience*, some of the highly focused success conditions are 'rigour of project design development' and 'proficiency of project team'. Within the viewpoint of successful/ failed project-based findings, a highly focused condition is 'client's motivation to achieve sustainable outcomes'. Within the viewpoint of *Suggestions for clients*, a highly focused condition is 'early engagement of project team'.

Overall Part 1, Part 2, and Part 3 of Figure 7.3 show that the identification of success conditions by interview participants is considerably determined by the context in which participants are based. Part 4 shows that the identification of success conditions by interview participants also depends on the viewpoint used in interviews for the inquiry of success conditions.

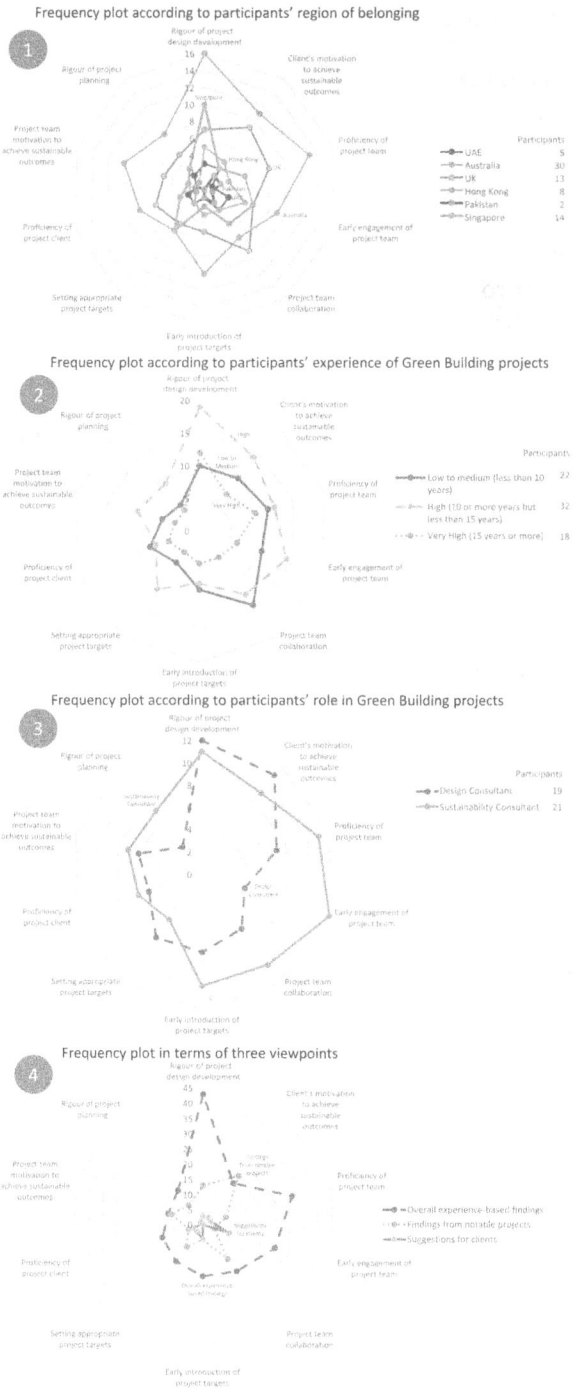

Figure 7.3 Effect of interview participants' demographic attributes and viewpoints on the identification of success conditions. (1) Frequency plot according to participants' region of belonging. (2) Frequency plot according to participants' experience of Green Building projects. (3) Frequency plot according to participants' role in Green Building projects. (4) Frequency plot in terms of three viewpoints.

7.3.1 Success conditions according to the role of interview participants in Green Building projects

Professionals belonging to different disciplines of the construction industry may associate different conditions with project success. Considering this, the semi-structured interviews in this study are conducted with professionals having different roles in GBs. The key findings of analysis with respect to two prominent roles of GB professionals (that is, design consultants and sustainability consultants) are as follows:

- Up to 19 interview participants having the role of design consultants in GBs discussed 53 conditions and 52 sub-conditions (Column J in Table 7.1). Up to 60 of these conditions and sub-conditions are identified by multiple interview participants. Some of the key conditions identified by more than six design consultants include 'Rigour of project design development' ($n = 12$), 'Client's motivation to achieve sustainable outcomes' ($n = 11$), 'Setting appropriate project targets' ($n = 7$), 'Proficiency of project team' ($n = 7$), and 'Early introduction of project targets' ($n = 7$).
- Up to 21 interview participants having the role of sustainability consultants in GBs discussed 53 conditions and 56 sub-conditions (Column K in Table 7.1). Up to 55 of these conditions and sub-conditions are identified by multiple interview participants. Some of the key conditions identified by more than nine sustainability consultants include 'Early engagement of project team' ($n = 12$), 'Rigour of project design development' ($n = 11$), 'Proficiency of project team' ($n = 11$), 'Project team collaboration' ($n = 10$), and 'Early introduction of project targets' ($n = 10$).

7.3.2 Success conditions according to the regional belonging of interview participants

The identified success conditions are analysed in terms of the six regions where the participants' experience of GB projects is mainly based. Following are the key findings of this analysis.

- Up to five interview participants having their experience of GBs in the UAE discussed 25 conditions and 17 sub-conditions (Column L in Table 7.1). Around 11 (26%) of these conditions are identified by multiple interview participants. Within the identified conditions, two conditions are unique for the UAE-based participants as they are not identified by participants from other regions. Some of the key conditions identified by more than two participants include 'Client's motivation to achieve sustainable outcomes' ($n = 3$), 'Rigour of project design development' ($n = 3$), 'Project team collaboration' ($n = 3$), and 'Early engagement of project team' ($n = 3$).

- Up to 30 interview participants having their experience of GBs in Australia discussed 62 conditions and 56 sub-conditions (Column M in Table 7.1). From among the conditions identified by Australia-based participants, eight conditions and 11 sub-conditions are not identified by participants based in other regions. Up to 76 conditions and sub-conditions are identified by multiple Australia-based participants. Some of the key conditions identified by more than nine participants include 'Rigour of project design development' ($n = 16$), 'Proficiency of project team' ($n = 13$), 'Client's motivation to achieve sustainable outcomes' ($n = 11$), 'Project team motivation to achieve sustainable outcomes' ($n = 10$), and 'Early introduction of project targets' ($n = 10$).
- Up to 13 interview participants having their experience of GBs in the UK discussed 50 conditions and 54 sub-conditions (Column N of Table 7.1), including three conditions and eight sub-conditions unique in terms of regional context. Up to 56 of these conditions and sub-conditions are identified by multiple interview participants. Some of the key conditions identified by more than six participants include 'Client's motivation to achieve sustainable outcomes' ($n = 9$), 'Project team collaboration' ($n = 9$), 'Proficiency of project team' ($n = 8$), and 'Rigour of project design development' ($n = 7$).
- Up to eight interview participants having their experience of GBs in Hong Kong discussed 32 conditions and 29 sub-conditions (Column O of Table 7.1), including two sub-conditions unique in terms of regional context. Up to 22 of these conditions and sub-conditions are identified by multiple interview participants. Some of the key conditions identified by more than three participants include 'Rigour of project design development' ($n = 5$), 'Early engagement of project team' ($n = 5$), and 'Client's motivation to achieve sustainable outcomes' ($n = 4$).
- Two interview participants having their experience of GBs in Pakistan discussed 14 conditions and 11 sub-conditions (Column P of Table 7.1), including one sub-condition unique in terms of regional context. Up to three of these conditions are identified by multiple interview participants. These conditions include 'Client's motivation to achieve sustainable outcomes' ($n = 2$), 'Proficiency of project team' ($n = 2$), and 'Project team collaboration' ($n = 2$).
- A total of 14 interview participants having their experience of GBs in Singapore discussed 39 conditions and 36 sub-conditions (Column Q of Table 7.1), including two conditions and one sub-condition unique in terms of regional context. Up to 35 of these conditions and sub-conditions are identified by multiple interview participants. Some of the key conditions identified by more than four participants include 'Rigour of project design development' ($n = 10$), 'Setting appropriate project targets' ($n = 6$), 'Early engagement of project team' ($n = 6$), 'Proficiency of project team' ($n = 5$), and 'Project team collaboration' ($n = 5$).

7.3.3 Success conditions according to the professional experience of interview participants

Three groups are created to analyse the identified success conditions in terms of the interview participants' professional experience of GB projects (that is, in number of years they have worked on GBs). This analysis can help see the effect of participants' professional experience on the identification of success conditions. There are 22 participants with low- to medium-level experience (that is, less than ten years), 32 participants with high-level experience (that is, more than ten but less than 15 years), and 18 participants with very high-level experience (that is, 15 years or more). Following are the key findings of this analysis.

- Participants with low- to medium-level experience discussed 54 conditions and 58 sub-conditions (Column R in Table 7.1). Up to 56 of these conditions and sub-conditions are identified by multiple interview participants. Some of the key conditions identified by more than nine participants include 'Project team collaboration' ($n = 14$), 'Proficiency of project team' ($n = 11$), 'Client's motivation to achieve sustainable outcomes' ($n = 10$), 'Rigour of project design development' ($n = 10$), and 'Early engagement of project team' ($n = 10$).
- Participants with high-level experience discussed 61 conditions and 67 sub-conditions (Column S in Table 7.1). Up to 80 of these conditions and sub-conditions are identified by multiple interview participants. Some of the key conditions identified by more than 11 participants include 'Rigour of project design development' ($n = 19$), 'Client's motivation to achieve sustainable outcomes' ($n = 14$), 'Early engagement of project team' ($n = 14$), 'Proficiency of project team' ($n = 12$), and 'Project team collaboration' ($n = 12$).
- Participants with very high-level experience discussed 55 conditions and 49 sub-conditions (Column T in Table 7.1). Up to 52 of these conditions and sub-conditions are identified by multiple interview participants. Some of the key conditions identified by more than five participants include 'Rigour of project design development' ($n = 12$), 'Proficiency of project team' ($n = 9$), 'Client's motivation to achieve sustainable outcomes' ($n = 7$), and 'Early engagement of project team' ($n = 6$).

It is important to notice that 'Client's motivation to achieve sustainable outcomes' is among the top three conditions highlighted by all three groups. Moreover, 'Early engagement of project team' is among the top five conditions highlighted by all three groups (that is, participants with low to medium, high, and very high experience of GB development).

7.4 Analysis of success conditions based on the viewpoints of interview participants

Three viewpoints are used by interview participants to discuss the conditions associated with GB project success, which are *Findings from successful/failed projects*, *Findings from participants' overall experience*, and *Suggestions for clients*. As shown in Figure 7.4, the same condition can occur within all three viewpoints. Figure 7.4 presents the case of a condition, 'Setting appropriate project targets', identified by 21 interview participants overall. From the viewpoint of *Findings from successful/failed projects*, this condition was associated with challenges ($n = 1$), failure ($n = 2$), and success ($n = 6$). From the viewpoint of the *Findings from participants' overall experience*, this condition was associated with operational failure ($n = 1$) as well as GB development success ($n = 10$) and failure ($n = 4$). This condition was also identified from the viewpoint of *Suggestions for clients* as three participants suggested this condition for the GB project clients. The reason that a condition is associated with both the success and failure of GB projects is that some states of the condition contribute towards success while some other states contribute towards failure as shown in Figure 7.4. While 69 (44%) of the overall conditions are only associated with project success, 11 (7%) of the overall conditions are only associated with project failure (as shown in Table 7.2). Hence, 75 (48%) conditions are associated with both the project success and failure, and this implies that for a sizable number of conditions the variable states of conditions can lead to different performance outcomes (that is, success and failure).

As many as 64 (41%) of all conditions are identified by only one out of the three viewpoints. Following is a brief overview of these conditions:

- Up to 41 (26%) of the overall 155 conditions are only identified when using the viewpoint of the *Findings from participants' overall experience* and are not identified through the other two viewpoints (Table 7.2). Some of these conditions identified by more than eight

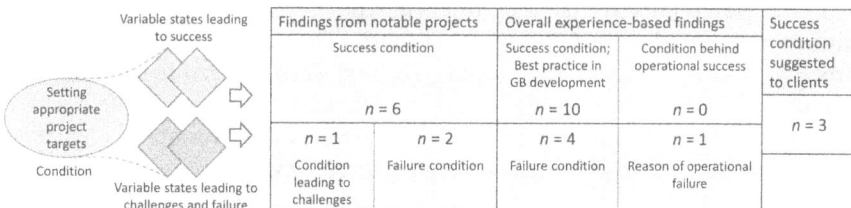

Variable states leading to success	Findings from notable projects		Overall experience-based findings		Success condition suggested to clients
	Success condition		Success condition; Best practice in GB development	Condition behind operational success	
	$n = 6$		$n = 10$	$n = 0$	$n = 3$
	$n = 1$	$n = 2$	$n = 4$	$n = 1$	
	Condition leading to challenges	Failure condition	Failure condition	Reason of operational failure	

Figure 7.4 Example of a success condition responding to the viewpoint of 'Successful/failed projects', 'Overall experience', and 'Suggestions for clients'.

participants include 'Educating end-users and FM team about building operation' ($n = 9$), 'Alignment of team interest with project interest' ($n = 9$), and 'Monitoring of project development' ($n = 12$).

- Up to 13 conditions and eight sub-conditions are only identified when using the viewpoint of successful/failed projects and are not identified through the other two viewpoints (Table 7.2). A highly identified condition in this regard is 'Structure and nature of client organisation' ($n = 5$). A highly identified sub-condition in this regard is 'PM team motivation to achieve sustainable outcomes' ($n = 3$).
- Two sub-conditions are only identified when using the viewpoint of *Suggestions for clients* and are not identified through the other two viewpoints (Table 7.2). These are 'Client's facilitation of coordination among project team' ($n = 1$) and 'FM team involvement in defining project aspirations' ($n = 1$).

This indicates that a sizable number of conditions (that is, 41%) are easily recalled by interview participants when employing multiple viewpoints in interviews. This implies that the use of multiple viewpoints for a topic of interest (such as GB success conditions) during interviews can ensure a richer data collection. While the use of multiple viewpoints in interviews contributed towards a comprehensive list of success conditions, it also resulted in data triangulation. A majority (that is, 59%) of conditions is identified by multiple viewpoints, and this indicates that the majority of conditions is triangulated by multiple perspectives. A detailed analysis regarding the overlap of success conditions among three viewpoints is shown in Figure 7.5.

Overall, the use of multiple viewpoints helped identify a relatively large number of success conditions per participant. By asking the interview participants to use multiple viewpoints during interviews, a better acquisition of required information was enabled since multiple viewpoints helped avoid the availability heuristic and bandwagon effect. The use of multiple viewpoints has also resulted in the triangulation of findings and a detailed understanding of success conditions.

Three viewpoints are used by interview participants to discuss the conditions associated with GB project success. The three viewpoints (which are *Findings from successful/failed projects, Findings from participants' overall experience,* and *Suggestions for clients*) shown in Table 7.1 are exclusively visualised in Figure 7.5 which depicts the overlap of identified conditions and sub-conditions across the three viewpoints.

From Part 1 of Figure 7.5, it can be seen that the viewpoint of 'successful/ failed projects' has conditions with three paradigms, that is (1) they are success leading, (2) failure leading, and (3) challenge leading. Within this viewpoint, 35 conditions and 33 sub-conditions belong to only one out of the three paradigms; 19 conditions and 11 sub-conditions belong to at least

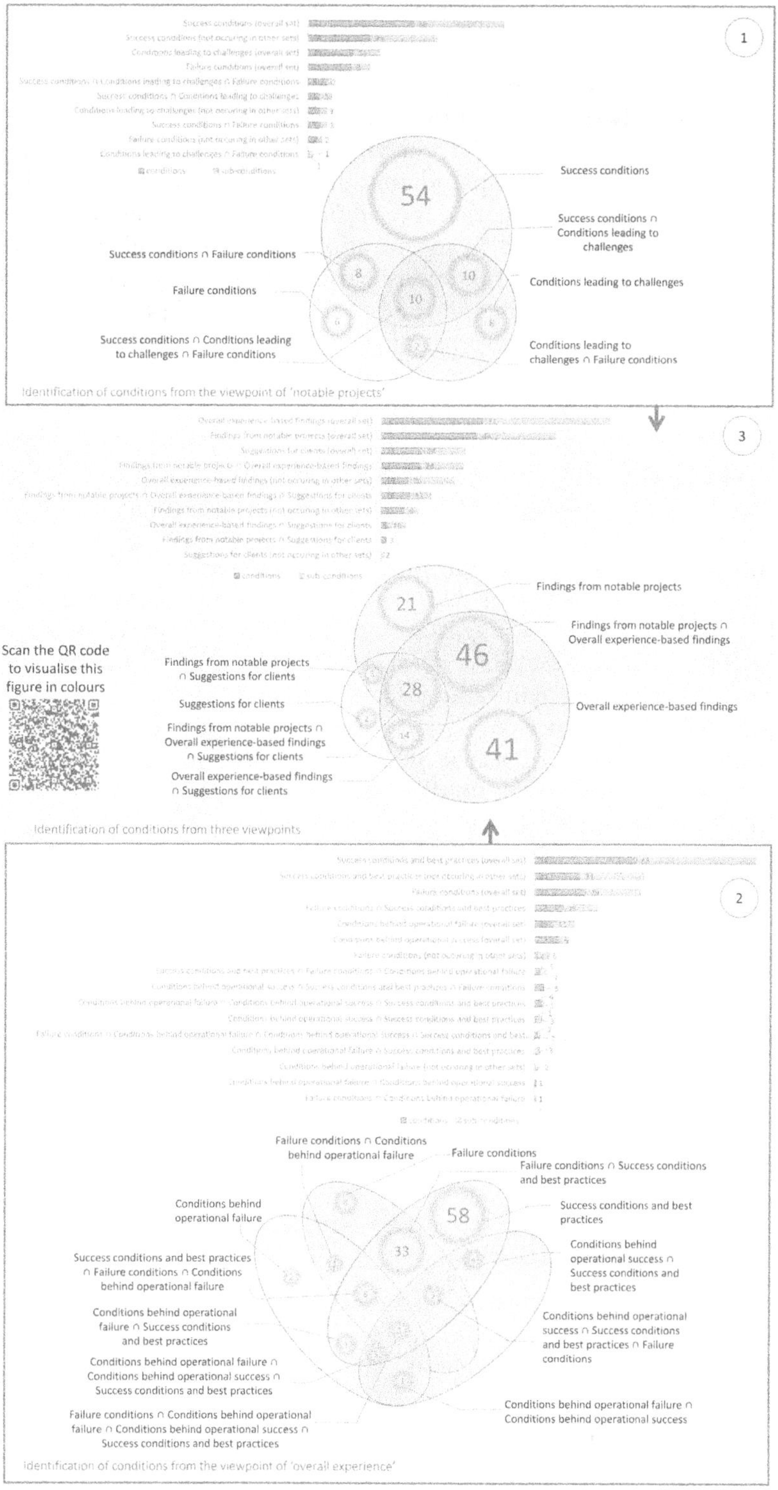

Figure 7.5 Overlap of conditions responding to the viewpoint of 'Successful/failed projects', 'Overall experience', and 'Suggestions for clients'.

two out of the three paradigms; and eight conditions and two sub-conditions concurrently belong to all the three paradigms.

From the Part 2 of Figure 7.5, the viewpoint of 'overall experience' has conditions belonging to two paradigms. There are two types of conditions within the success paradigm, that is conditions leading to the overall project success and conditions leading to operational success. Similarly, there are two types of conditions within failure paradigm, that is conditions leading to the overall project failure and conditions leading to operational failure. In Figure 7.5 (Part 2), the overlap of the four types of conditions (that is, conditions for operational success, operational failure, overall success, and overall failure) is shown.

Part 3 of Figure 7.5 shows an overlap among the conditions identified within the three viewpoints (that is, *Findings from successful/failed projects*, *Findings from participants' overall experience*, and *Suggestions for clients*). Overall, the highest number of conditions (57) and sub-conditions (72) is identified within the viewpoint of 'overall experience'. Least number of conditions (23) and sub-conditions (24) is identified within the viewpoint of *Suggestions for clients*. This may be owing to the fact that in response to this viewpoint, the interview participants in their discussion were mainly limited to the conditions controllable by a project client. In the case of *Suggestions for clients* viewpoint, other than two sub-conditions, all the rest of 23 conditions and 22 sub-conditions also concurrently occur within the other two viewpoints. This points out that in case the other two viewpoints are used in interviews for identifying the success factors, the *Suggestions for clients* viewpoint has little contribution in developing an exhaustive dataset of conditions. However, the considerable overlap of this viewpoint with the other two viewpoints does not undermine the importance of this viewpoint for a qualitative inquiry since such an investigation is more than the mere identification of conditions. Overall, 45 conditions and 46 sub-conditions are identified in at least two out of the three viewpoints, and 28 conditions and 36 sub-conditions are exclusively identified in one of the three viewpoints. This highlights the importance of using multiple viewpoints in developing an exhaustive dataset of success conditions.

7.4.1 Viewpoint-1: Findings from the overall experience of interview participants

From the viewpoint of their overall experience of GB projects, interview participants have identified conditions and sub-conditions, leading to success and failure of GB project development as well as operation.

Up to 55 interview participants associated 106 conditions with the success of GB project development (shown in Column H of Table 7.1). Among these conditions, 72 (68%) are highlighted by multiple interview participants. Some of the conditions discussed by more than 11 participants include 'Rigour of project design development' (n = 21), 'Proficiency of project team' (n = 21), 'Project team collaboration' (n = 15), 'Early engagement of project team' (n = 14), and 'Client's motivation to achieve sustainable outcomes' (n = 12).

Up to 30 interview participants associated 56 conditions with the failure of GB project development (shown in Column D of Table 7.1). Among these conditions, 29 (52%) are highlighted by multiple interview participants. Some of the conditions discussed by more than five participants include 'Rigour of project design development' (n = 11), 'Proficiency of project team' (n = 7), 'Rigour of project planning' (n = 6), and 'Early introduction of project targets' (n = 6).

Up to eight interview participants associated 18 conditions with the operational success of GB projects (shown in Column F of Table 7.1). Some of these conditions discussed by multiple interview participants include 'Educating end-users and FM team about building operation' (n = 5), 'End-users' operation of building in sustainable ways' (n = 2), 'Smooth transition of project from inception to operation stage' (n = 2), 'Inspection of project upon construction' (n = 2), and 'Early engagement of project team' (n = 2).

Up to ten interview participants associated 21 conditions with the operational failure of GB projects (shown in Column E of Table 7.1). Some of these conditions discussed by multiple interview participants include 'Rigour of project design development' (n = 6), 'End-users' operation of building in sustainable ways' (n = 2), 'Educating end-users and FM team about building operation' (n = 2), 'Smooth transition of project from inception to operation stage' (n = 2), 'Monitoring and controlling operational performance of building' (n = 2), and 'Proficiency of FM team' (n = 2).

7.4.2 *Viewpoint-2: Findings from successful/failed projects*

The interview participants recalled some challenging, failed, and successful GB projects they have been involved in and highlighted the conditions and sub-conditions resulting in challenges (Column B), failure (Column C), and success (Column G) in those projects. For instance, two interview participants associated 'Project team's access to robust information' with challenges (Column B), three interview participants associated 'Scope changes during project execution' with GB project failure (Column C), and one interview participant associated 'Client's motivation to achieve original project intent' with project success (Column G).

While referring to the successful/failed projects in their professional careers, the interview participants associated the conditions and sub-conditions with success and failure, as well as challenging circumstances in GB projects. As shown in Figure 7.6, these conditions have originated from

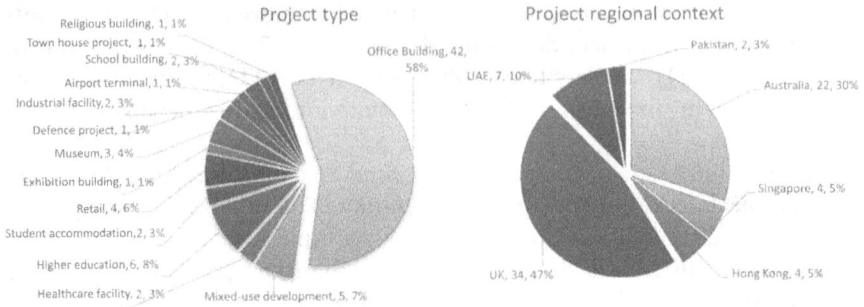

Figure 7.6 Functional use and regional context of successful/failed Green Building projects discussed in interviews.

73 GB projects having a variety of functions (such as office, mix-use, and retail) and spread across six regions, that is, the UAE (number of projects = 7), Australia (n = 22), the UK (n = 34), Hong Kong (n = 4), Pakistan (n = 2), and Singapore (n = 4). Most of the mentioned projects are office building projects (n = 42, 58%), while the rest of the projects (n = 31, 42%) have many different functional categories as shown in Figure 7.6. Further details regarding individual projects are provided in Table 7.3.

In the case of the successful/failed projects discussed in interviews, 19 conditions and 11 sub-conditions (shown in Column B of Table 7.1) are associated with challenging circumstances by 15 interview participants. The conditions associated with challenges by more than two interview participants are 'Rigour of project planning' (n = 4) and 'Stringency level of project sustainability requirements' (n = 3).

While discussing the successful/failed GB projects in their professional careers, nine interview participants associated 18 conditions and eight sub-conditions with project failure (shown in Column C of Table 7.1). The conditions associated with GB failure by more than one interview participant are 'Scope changes during project execution' (n = 3), 'Priority of sustainability in project development' (n = 2), 'Setting appropriate project targets' (n = 2), and 'Proficiency of project client' (n = 2).

While discussing the GB projects, 32 interview participants associated 44 conditions and 38 sub-conditions with project success (shown in Column G of Table 7.1). The conditions associated with GB project success by more than seven interview participants are 'Client's motivation to achieve sustainable outcomes' (n = 20), 'Project team collaboration' (n = 13), 'Rigour of project design development' (n = 11), and 'Project team motivation to achieve sustainable outcomes' (n = 8).

7.4.3 Viewpoint-3: Suggestions for clients

A client is the most influential project team member. Different actions enabled by the project client or through the client's representatives (for instance,

Table 7.3 Attributes of Green Building projects discussed by interview participants

Project ID	Country	Project functional use	Duration of project delivery (no. of years)	Project completion date (planned/actual)	Approximate project scope (covered floor area/cost, etc.)	Level of technical complexity			Overall performance			Sustainability performance			Interview participant ID	Role of participant/participant's company in the project
						Low	Medium	High	Success	Failure	Mix	Achieved aspirations	Achieved more than aspirations	Achieved less than aspirations		
AE-A-1	UAE	Airport terminal	4	2019	250,000 m²			x	x			x			AE-M-2	Sustainability consultant
AE-I-1	UAE	Industrial facility	2	2012	Overall covered floor area of complex = 11,000 m²		x		x			x			AE-M-4	Design consultant; sustainability consultant
AE-M-1	UAE	Museum	3	2019	5000 m²			x	x			x			AE-M-1	Sustainability consultant
AE-O-1	UAE	Office building	1.3	2016	80,000 ft²		x		x						AE-F-1	Client
AE-O-2	UAE	Office building	0.5	2018	400 m²		x		x						AE-F-1	Client
AE-O-3	UAE	Office building	2.5	2018	Three floors		x		x			x			AE-F-2	Design consultant; sustainability consultant
AU-DE-1	Australia	Defence project	6								x	x			AU-M-3	Sustainability consultant
AU-DO-1	Australia	Student accommodation	1.7	2018	150 beds			x			x	x			AU-M-8	Project manager
AU-E-1	Australia	Higher education facility	3	2017	28,000 m² GFA; four floors			x	x			x			AU-M-8	Project manager
AU-H-1	Australia	Healthcare facility	5	2017	8 acres site area										AU-F-7	Design consultant
AU-MU-1	Australia	Mixed-use development	1	2017	1300 m²				x			x			AU-F-7	Design consultant
AU-O-1	Australia	Office building	2	2009	17 storeys; 11 floor office tower @ 1650 m² of NLA per floor		x		x			x			AU-F-2	Design consultant

(Continued)

Table 7.3 Attributes of Green Building projects discussed by interview participants (Continued)

Project ID	Country	Project functional use	Duration of project delivery (no. of years)	Project completion date (planned/actual)	Approximate project scope (covered floor area/ cost, etc.)	Level of technical complexity			Overall performance			Sustainability performance			Interview participant ID	Role of participant/ participant's company in the project
						Low	Medium	High	Success	Failure	Mix	Achieved aspirations	Achieved more than aspirations	Achieved less than aspirations		
AU-O-10	Australia	Office building		2016	$200 Million; 32 floors		X					X			AU-M-9	Head contractor
AU-O-11	Australia	Office building	1.5	2016			X					X			AU-F-8	Design consultant
AU-O-12	Australia	Office building					X			X		X			AU-F-8	Sustainability certification assessor
AU-O-13	Australia	Office building	4	2010	16,000 m²				X			X			HK-M-4	Design consultant
AU-O-2	Australia	Office building	5	2004	3000 m²		X		X				X		AU-M-3	Sustainability consultant
AU-O-3	Australia	Office building									X	X			AU-M-3	MEP consultant; sustainability consultant
AU-O-4	Australia	Office building			15 Floors; 27,850 m² NLA				X				X		AU-F-3	MEP consultant; sustainability consultant
AU-O-5	Australia	Office building			5000 m²; two floors					X				X	AU-F-3	Sustainability consultant
AU-O-6	Australia	Office building	3		27 floors	X			X			X			AU-F-4	MEP consultant; Sustainability consultant
AU-O-7	Australia	Mixed-use development	3	Ongoing	15,000 m²; 2 floors				X				X		AU-F-6	MEP consultant; Sustainability consultant
AU-O-8	Australia	Office building	2	2013	63,000 m²; $320 Million			X	X			X			AU-M-7	Head contractor
AU-O-9	Australia	Mixed-use development	3	2006	Three floors; $40 Million			X					X		AU-M-8	Project manager
AU-R-1	Australia	Town house project	6	2021	150 housing units						X			X	AU-M-1	Developer; design consultant; head contractor
AU-S-1	Australia	Retail			$600 Million				X			X			AU-M-4	Design consultant

Code	Country	Building type		Year	Size	Participant	Role
AU-S-2	Australia	Retail	5	2019	12,500 m²	AU-M-6	Client
AU-S-3	Australia	Retail		2019	6000 m²	AU-M-10	Sustainability consultant
HK-EX-1	Hong Kong	Exhibition building		2013		HK-M-7	Design consultant
HK-HE-1	Hong Kong	Higher education facility				HK-M-7	Design consultant
HK-HE-2	Hong Kong	Higher education facility			GFA 45,000 m²	HK-M-2	Sustainability consultant
HK-O-1	Hong Kong	Office building			13 floors	HK-M-2	Sustainability consultant
PK-I-1	Pakistan	Industrial facility	1.5	2017	50,500 ft²; 500 Million PKR	PK-M-1	Design consultant; MEP consultant; sustainability consultant
PK-O-1	Pakistan	Office building	1.1	2018	50,000 ft²; 5 Million USD	PK-M-2	Head contractor
SN-H-1	Singapore	Healthcare facility	Ongoing	2020	250,000 m²	SN-F-2	Sustainability consultant
SN-HE-1	Singapore	Higher education facility	Ongoing			SN-F-1	Client; building user
SN-MU-1	Singapore	Mixed-use development				HK-M-4	Design consultant
SN-O-1	Singapore	Office building	3	2016	GFA: 4.5 million	SN-F-2	Design consultant; sustainability consultant
UK-DO-1	UK	Student accommodation	1.5	2016	26 floors	UK-F-2	Sustainability consultant; sustainability certification assessor; renewable feasibility consultant; post-occupancy assessor
UK-E-1	UK	School building	6	2017	2000 m²	UK-F-4	Sustainability certification assessor
UK-E-2	UK	School building	2	2015		UK-M-1; UK-M-2	Sustainability consultant
UK-HE-1	UK	Higher education facility	4	2017		UK-F-6	Sustainability consultant
UK-HE-2	UK	Higher education facility	20	2020	2 B£	UK-M-6	Sustainability consultant
UK-MU-1	UK	Mixed-use development	6	2014	85 M£	UK-F-1	Energy consultant; sustainability consultant

(Continued)

Table 7.3 Attributes of Green Building projects discussed by interview participants (Continued)

Project ID	Country	Project functional use	Duration of project delivery (no. of years)	Project completion date (planned/actual)	Approximate project scope (covered floor area/cost, etc.)	Level of technical complexity			Overall performance			Sustainability performance			Interview participant ID	Role of participant/participant's company in the project
						Low	Medium	High	Success	Failure	Mix	Achieved aspirations	Achieved more than aspirations	Achieved less than aspirations		
UK-O-1	UK	Office building	1.5	2017	Five floors; 15,000 m²		×				×	×			UK-F-1	Sustainability certification assessor
UK-O-10	UK	Office building	2.5	2017	40 M£			×	×					×	UK-M-7	Head contractor
UK-O-11	UK	Office building	3	2018	70 M£			×		×		×			UK-M-7	Head contractor
UK-O-12	UK	Office building	1	2015	10 M£	×			×			×			UK-M-7	Sustainability consultant
UK-O-2	UK	Office building	1.5	2016	Six floors; 3,000 m²		×		×				×		UK-F-2	Sustainability consultant; BREEAM assessor; renewable feasibility consultant
UK-O-3	UK	Office building					×			×		×			UK-F-3	
UK-O-4	UK	Office building	3	2019	10,000 m²; 70 M£		×		×			×			UK-F-4	Engineering consultant (structure, civil, VT, acoustics, MEP, and sustainability)
UK-O-5	UK	Office building	7	2017	5000 m²		×		×			×			UK-F-5	Design consultant
UK-O-6	UK	Office building	2	2015				×	×			×			UK-F-6	Sustainability consultant
UK-O-7	UK	Office building	3	2017			×		×			×			UK-F-6	Sustainability consultant
UK-O-8	UK	Office building	5	2017	625 M£			×	×			×			UK-M-5	Sustainability consultant
UK-O-9	UK	Office building	5	2017	615 M£	×				×		×			UK-M-5	Sustainability certification assessor
UK-R-1	UK	Religious building	8	2016			×			×				×	UK-M-1; UK-M-2	Sustainability consultant
UK-SC-1	UK	Retail	11		500 M£		×		×			×			UK-F-3	Sustainability consultant

project manager or architect) can make a substantial difference in the project performance. The interview participants were requested to provide their suggestions from the standpoints of sustainability experts to a fictional client for the development of a GB project.

While discussing the *Suggestions for clients*, 19 interview participants associated 23 conditions and 24 sub-conditions with project success (shown in Column I of Table 7.1). Up to 23 of these conditions and sub-conditions are highlighted by multiple interview participants. Some of these conditions highlighted by more than three participants include 'Early engagement of project team' ($n = 9$), 'Early introduction of project targets' ($n = 5$), and 'Proficiency of project client' ($n = 4$).

The suggestions for a client are not merely applicable in parts. Instead, it is important that the client observes these suggestions collectively and in a timely manner.

According to an Australia-based interview participant (AU-F-5),

> The client needs to finalise the level of the certification; the project will strive for. Afterwards, the client needs to discuss with a sustainability consultant the credits in the certification system which are the target for the new development. Once the intention of credit is decided, it needs to be documented. This is because on knowing the credits to target, the design consultant is able to design a building accordingly. In case the client adds something in the program of development after tendering is done, then it starts costing money.

7.5 Summary

Epistemic relativism used in this study indicates that knowledge is articulated from various standpoints according to various influences and interests and is transformed by human activity. For the inquiry of GB project success, the use of epistemic relativism implies that industry experts with differing role and experience in GB projects and different regional belonging may perceive the success in GBs differently. This chapter provides additional analysis to explore the effect of the attributes and viewpoints of interview participants on the identified success conditions.

In semi-structured interviews, multiple viewpoints were employed by industry experts when talking about GB success conditions. Some of the prevalent viewpoints include 'findings from overall experience of working on GB projects', 'findings from particular successful/failed projects', and 'suggestions for future clients of GB projects'. The detailed analyses indicate that the identification of success conditions is affected by the viewpoints employed by interview participants when thinking and talking about GB projects and also by the regional belonging, experience, and role of interview participants in GB projects.

Index

For Product Safety Concerns and Information please contact our EU
representative GPSR@taylorandfrancis.com
Taylor & Francis Verlag GmbH, Kaufingerstraße 24, 80331 München, Germany